HOW TO BUILD Max-Performance PONTIAC V-8s

Rocky Rotella

CarTech®

CarTech®

CarTech®, Inc.
838 Lake Street South
Forest Lake, MN 55025
Phone: 651-277-1200 or 800-551-4754
Fax: 651-277-1203
www.cartechbooks.com

© 2012 by James "Rocky" Rotella

All rights reserved. No part of this publication may be reproduced or utilized in any form or by any means, electronic or mechanical, including photocopying, recording, or by any information storage and retrieval system, without prior permission from the Publisher. All text, photographs, and artwork are the property of the Author unless otherwise noted or credited.

The information in this work is true and complete to the best of our knowledge. However, all information is presented without any guarantee on the part of the Author or Publisher, who also disclaim any liability incurred in connection with the use of the information.

All trademarks, trade names, model names and numbers, and other product designations referred to herein are the property of their respective owners and are used solely for identification purposes. This work is a publication of CarTech, Inc., and has not been licensed, approved, sponsored, or endorsed by any other person or entity.

Edit by Scott Parkhurst
Layout by Sue Luehring

ISBN 978-1-61325-474-5
Item No. SA233P

Library of Congress Cataloging-in-Publication Data

Rotella, Rocky.
  How to build max-performance Pontiac V-8s / by Rocky Rotella.
    p. cm.
  Based on 2004 version by Jim Hand.
  ISBN 978-1-934709-94-8
  1. Pontiac automobile–Motors–Modification–Handbooks, manuals, etc. 2. Pontiacautomobile–Performance–Handbooks, manuals, etc. I. Hand, Jim. How to build max-performance Pontiac V-8s. II. Title.

TL215.P68R635 2012
629.25–dc23

2012028447

Written, edited, designed and printed in the U.S.A.

Title Page: *Certain cylinder heads have an intake port arrangement that requires a specific intake manifold design. Some manufacturers offer matching cast-aluminum intake manifolds, but others like this unit from KRE (for its Warp-6 heads), must be custom fabricated for the exact application. As long as hood clearance isn't a concern, the possibilities are limitless, including one or more carburetors.*

Back Cover Photos

Top Left: *There are a few different aftermarket Pontiac V-8 blocks presently available that are ideal for applications approaching 750 hp and beyond. AllPontiac.com manufactures two distinct units. The IA-II is a cast-iron block that's been on the market for several years while the all-new aluminum block was a recent release. Both generally boast beefier main saddles and four-bolt main caps to significantly improve bottom end rigidity, particularly when using a very long stroke aftermarket crankshaft measuring 4.5 inches or more to produce a maximum displacement of 541 ci.*

Top Right: *The KRE High Port was designed as a modern replacement for the Pontiac round port. Peak intake airflow measures 330 cfm as cast, but optional porting increases that toward 400 cfm. It uses a typical intake manifold and round-port exhaust headers.*

Middle Left: *Crankshaft thrust is taken up on the number-4 main journal. Whether using an original Pontiac crank or one of the main aftermarket options, thrust clearance should remain between .005 and .009 inch. Any variance outside that range can lead to premature bearing wear and/or abnormal operating issues. Selecting main bearings from another manufacturer can sometimes affect it, but I suggest contacting your machinist and component suppliers if it varies outside that range.*

Middle Right: *A camshaft degree kit, such as this one from Comp Cams, is an excellent tool that any hobbyist should learn to use. It's used to accurately measure and/or calculate all the critical valve events of a particular camshaft and to ensure that it's installed at the desired crankshaft angle. Verifying the installation point provides maximum possible performance, especially if slight adjustment is required.*

Bottom Left: *Before placing any connecting rod in service, have your local machinist verify that the crank and wrist pin bores measure within the specified tolerances. In most instances you should find nothing major. Your machinist can easily hone them to the required specification if any small variance is found.*

Bottom Right: *The Northwind manifold from KRE has a large plenum and runners, but its carburetor flange is only slightly taller than stock. It provides excellent performance without significant hood clearance issues and plenty of room for porting. It's another excellent choice for competition engines.*

CONTENTS

Preface 4
Acknowledgments 5
Foreword by Thomas A. DeMauro 6

Chapter 1: History of the Pontiac V-8 1955–1981 7
V-8 Development 7
The V-8 Reaches Production 7
Growth Spurts 8
The Super Duty Era 9
The Corporate Edict 10
DeLorean Promoted 10
Major V-8 Design Changes 11
The Round-Port Era 12
Living with Low Compression 12
The Second Super Duty Era 13
Emissions and Economy 14
Emissions-Compliant Performance 14
The End of an Era 14
Pontiac Racing 14

Chapter 2: Blocks 16
Block Design 16
Selecting a Stock Block 17
Stock Block Modifications 19
Water Jacket Filler 21
Main Caps 22
Aftermarket Blocks 24
Camshaft Bearings 26
Rear Main Seal 26
Fasteners 28
Head Gaskets 30
Bolt-On Components 31

Chapter 3: Crankshafts 33
Original Pontiac Cranks 33
Selecting a Stock Crank 34
Original Crank Modifications 35
Stroker Cranks 37
Aftermarket Cranks 38
Specialty Billet Cranks 40
Balancing 41
Harmonic Dampers 41
Engine Bearings 43

Chapter 4: Connecting Rods 46
Original Pontiac Rods 46
Aftermarket Steel Rods 47
Aluminum Rods 49
Connecting Rod Length 51
Fasteners 51
Selection, Preparation and Inspection ...52

Chapter 5: Pistons and Rings 53
Pistons 53
Piston Rings 56
Preparation and Inspection 58
Piston Coatings 59

Chapter 6: Cylinder Heads 60
Intake Airflow 60
Port Volume 61
Exhaust Airflow 61
High-Performance Valve Job 63
Combustion Chamber 63
Compression Ratio 64
Porting Advice 65
Cylinder Head Options 65

Chapter 7: Valvetrain 78
Hydraulic Lifters 78
Factory Cams 79
Factory Rocker Arms 81
Aftermarket Hydraulic Cams 82
Aftermarket Solid Cams 82
Aftermarket Roller Cams 84
Aftermarket Rocker Arms 86
Aftermarket Rocker Studs 88
Aftermarket Pushrods 89
Valvesprings and Valves 90

Chapter 8: Intake Manifolds 92
Factory Cast-Iron 4-Barrel 93
Factory High-Performance 4-Barrel 94
Factory Tri-Power 94
Factory Reproductions 95
Selecting a Factory 4-Barrel 95
Modifying Factory Manifolds 96
Aftermarket Manifolds 97
Carburetors 100
Fuel Injection 102
Forced Induction 104

Chapter 9: Exhaust 105
High-Performance Exhaust Manifolds ...106
Reproduction Factory Manifolds 107
Tubular Headers 107
Ceramic Coating 110
X- and H-Type Crossovers 110
Mufflers 111
Complete Exhaust Systems 111

Chapter 10: Ignition 112
Spark Lead 112
Points-Type System 113
Electronic Ignition System 114
Aftermarket Ignition System 115
Ignition Accessories 116

Chapter 11: Oiling System 118
Oil 118
Oil Pump 120
Oil Pressure Requirements 121
Oil Filter 122
Oil Pan 123

Chapter 12: Tuning 125
Carburetor Tuning 125
Distributor Tuning 128
Dyno Tuning 129
Volumetric Efficiency 130
Vacuum Pump 131
Nitrous Oxide 131
Flywheel and Flexplate 132
Water Pump 133
Starter 133

Chapter 13: Performance Combinations 134
Pacific Performance Racing: 670 hp, 383 ci, Pump Gas 135
Ken's Speed & Machine Shop: 700 hp, 535 ci, Pump Gas 136
SD Performance: 900 hp, 535 ci, Race Gas 137
Butler Performance: 1,000+ hp, Turbocharged 505 ci, Pump Gas 138
Butler Performance: 1,100+ hp, 541 ci, Race Gas 139
Kauffman Racing Equipment: 1,200 hp, 535 ci, Race Gas 140

Source Guide 141

PREFACE

Welcome to *How To Build Max-Performance Pontiac V-8s*. Right now you're probably saying, "Wait a minute. I think I've read this book before." Well, you're probably right, but hold on! This content is very different.

CarTech published a book with this very title in 2004. Noted Pontiac racer Jim Hand authored it and it delved into attaining maximum performance using mostly original Pontiac components, which at the time it was written, was about all that was readily available and affordable. It was very informative and an excellent read.

How times have changed. While very stout performance can still be attained using factory equipment, most modern high-performance Pontiac engines are built using very few original Pontiac pieces. It seems it's simply more convenient to use new components as opposed to salvaging old ones. In fact, most everything you need is available new, and not a single original component is required to build an entirely new Pontiac engine.

This book lists the best factory components as well as many of the most popular aftermarket pieces. There are a great number of proven combinations assembled by professional Pontiac engine builders. You're certain to find one that fulfills your performance expectations, and there's a countless number of directions you can take it to fit your particular needs.

If you're reading this book, you're likely familiar with the process required to successfully assemble a Pontiac V-8. If not, my CarTech book, *How to Rebuild Pontiac V-8s*, includes step-by-step instructions that explain the pre-assembly and assembly processes in great detail. It is a worthwhile investment if you have limited experience and plan to assemble your max-performance engine yourself.

I highly suggest finding a professional Pontiac engine builder you trust and are comfortable doing business with. A list of the most respected Pontiac builders can be found in the Source Guide at the end of this book. State your performance goals and budget, ask what the builder recommends, and purchase the equipment and components from them. It's the best way to ensure that you'll end up with the Pontiac performance you expect!

ACKNOWLEDGMENTS

There are a great number of people who have helped me further my Pontiac knowledge along the way. Whether local friends, Pontiac professionals, or even those faceless names I know through email or web forums, each of you deserves credit. I do, however, want to take a moment to recognize three individuals to whom I owe the sincerest gratitude for the education they've given me over the years. Each is a true Pontiac enthusiast whose name you'll most likely recognize. Without their guidance and belief, a project such as this would never have been possible.

Nunzi Romano is a Pontiac performance legend. For more than 40 years he owned and operated Nunzi's Automotive in Brooklyn, New York, building and racing some of the fastest Pontiacs in the country, using mostly factory-issued equipment. I grew up reading about Nunzi in *High Performance Pontiac* magazine (and its predecessor, *Thunder Am*). I first contacted Nunzi in the early 1990s and a friendship developed. I always listened as he shared his knowledge and explained performance theories. He has since retired, but hasn't slowed down any, now working on his own Pontiac projects. Without Nunzi's inspiration, I doubt that I'd ever be in a position to write about Pontiac performance. Thanks, Nunzi.

Jim Hand has been racing Pontiacs since the 1950s. He authored articles for *High Performance Pontiac* and *Pontiac Enthusiast* magazines. I was always in awe of his practical approach and ability to make great power using mostly stock equipment. I met Jim in 2000 and we began speaking regularly. Jim always encouraged me to test various components to gain firsthand experience on their effects, good or bad. He also encouraged me to purchase a distributor tester and flow bench, both of which I use regularly today. As a former author for CarTech, Jim suggested me as a possible author for Pontiac topics, and because of him, I've written my second published book with CarTech. Thank you, Jim.

Tom DeMauro is the current Editor In Chief at *High Performance Pontiac* magazine. I read the publication growing up and always thought about writing an article someday, but considered myself unqualified as my English class grades were average at best and I had no photographic skills. Tom had asked for reader assistance with a technical question he'd received. I read it and replied, and my response ran in a subsequent issue. I met Tom at a Pontiac event in 2001 and he invited me to submit an article. I initially declined due to lack of confidence, but Tom persisted. Whatever he saw that day has turned into more than 100 published articles so far. Without Tom's belief, and his literary and photographic guidance, I'm absolutely certain this book wouldn't be possible. Thank you, Tom.

I also wish to thank my father, Jim Rotella, for it was his enthusiasm toward Pontiacs that led me to literally grow up in and around them. We still regularly work together on Pontiac projects as well as attend large Pontiac events around the country. Last, but not least, my wife, Jennifer, has also been very supportive and tolerant of this project. It's taken away from valuable family time, especially with our newborn daughter, Sofia, arriving midway through the project, but she understands its importance and has been great through it all. She's the best wife a car guy could ever have.

There are many people who assisted in the compilation of information for this book. I couldn't have completed the project without the assistance of many people. I'd like to extend thanks to the following:

Toby Aldrete, Ken Anderson, Bruce Baldwin, Matt Barkhaus, Richard Batchelor, Will Baty, Joel Bayliss, Kevin Beal, Jeff Behuniak, Dave Bisschop, Ken Brewer, Armin Brown, Mike Burchell, Lane Burnett, David Butler, Mike Cameron, Brian Carson, Wade Congdon, Ken Crocie, Alan Davis, Paul Delfield, Tom DeMauro, Dave Hall, Brian Ellis, Bob Florine, Joe Flower, Scott Gabrielson, Dick Glady, Trent Goodwin, Frank Gostyla, Jim Hairston, Brandon Hamm, Jay Hancock, Jim Hand, Tom Hand, Ryan Hunter, Jack Isom, Terry Johnson, Jeff Kauffman, Don Keefe, Ken Keefer, Mark Lewis, Tom Lieb, Lou Lobsinger, Robert Loftis, Robert Martin, Bill McKnight, Don Meziere, Sheldon Miller, Randy Moore, Kerry Novak, Mike Osterhaus, Gary Otto, Chris Ouellette, Scott Parkhurst, Chris Phillip, Stan Poff, Dan Qualls, Scott Ray, Stan Ray, Nunzi Romano, Tony Romano, Jim Rotella, Ron Rotunno, Cliff Ruggles, Todd Ryden, Jim Sammons, Steve Schappaugh, Brian Scollon, Ann Skrycki, Scott Sims, Ken Sink, Doug Smith, Smitty Smith, Kevin Studaker, Zeke Urritia, Jim Wangers, Susan Weimar, Chuck Willard, Terry Wilson, and Kris Zdral.

FOREWORD

By Thomas A. DeMauro, Editor, *High Performance Pontiac* magazine

Back in the early 1990s when I started with *High Performance Pontiac* magazine, there were no aluminum heads, forged cranks, or engine blocks available in the aftermarket for the Pontiac V-8. Like most engines out of production for more than a decade, many of the existing speed parts dated to the heyday of its original-equipment (OE) usage.

Within a few years, however, companies realized that the Pontiac V-8 performance market had sales potential. Not just racers wanted modern, lighter, and better performing engine parts to go fast, but so did hobbyists who were building street cars.

As a result, the mid 1990s brought fresh intake manifold designs and aluminum heads, cast stroker cranks, and readily available forged-steel connecting rods. Aftermarket cylinder blocks to build big-bore, long-stroke monster motors soon followed. In the 2000s the market exploded with more of everything—blocks, intakes, new forged stroker cranks and stroker rotating assemblies, port and throttle-body EFI, and a vast array of cylinder heads.

Large and small companies, including engine builders, have designed and produced cylinder heads from round-port (even R/A V) and D-port types with modern chambers and revised ports that can still be used with stock parts, to high-ports, wide-ports, pro-port race heads for the builder to finish, Pro/Stock-style heads, and even canted-valve heads for Pontiacs.

The origin of this onslaught of Pontiac performance parts lies in the fact that the division designed a great OHV V-8 engine in the 1950s and kept refining it over its two-and-a-half decades of production. It also enjoys plentiful parts interchange from one displacement to another thanks to retaining a single overall block and head design with 3.00- or 3.25-inch mains and round-port or D-port heads being the most notable differences.

Tri-Power, Super-Duty, Ram Air, H.O, sound engineering and competitive power output combined with masterful marketing, ensured the Pontiac V-8 left an indelible mark in the minds of car lovers everywhere through the 1960s and 1970s. And the legend did not fade over the years that followed its demise.

The second wave of interest began in the late 1980s and early 1990s when baby boomers wanted to recapture their youth in a GTO, Firebird, T/A, GP, or even a big 2+2 or Bonneville.

Also came the children of the baby boomers, many of whom (including myself) had grown up in the backseat of Pontiacs. We too sought the power, performance, and the smoothness and refinement of a Pontiac when we came of driving age. For me, it led to the ownership of a 1966 GTO, a 1967 GTO, a 1969 Judge, a 1977 Trans Am, and a 1979 Formula over the years. I still own the 1967 GTO and 1977 T/A. Ours was a whole new generation for the speed parts manufacturers to satisfy.

Rocky Rotella also grew up in the backseat of some of the greatest Pontiacs ever made, thanks to the discerning automotive taste of his father, James. Rocky has been immersed in the Pontiac hobby seemingly since birth, and his enthusiasm for learning more about them never wanes. He has been a regular contributor to *High Performance Pontiac* magazine since 2003. In this book, Rocky's research and hands-on experience with the Pontiac V-8 and the aftermarket parts to build them will provide you with knowledge, while his writing skill, earned from crafting insightful stories for car magazines for nearly a decade, will entertain you in the process.

Today, sales of vintage Pontiac V-8-powered cars continue to move forward. The availability of well-engineered components to build a variety of unique and powerful engine combos—never dreamt of by PMD engineers back in the day—is at an all-time high and will ensure that those Pontiacs remain competitive on the street, the strip, and the road course.

Proof of the enduring popularity of the Pontiac V-8 is that this book is being published in 2012 for an (standard-deck) engine that hasn't been installed in a new vehicle since 1979. So pat yourself on the back for continuing to support our hobby. Then start reading. You and your Pontiac will certainly benefit.

CHAPTER 1

HISTORY OF THE PONTIAC V-8 1955–1981

The Oakland Motor Car Company was founded in Pontiac, Michigan, in the early 1900s, and within a few years, it became a division of the newly formed General Motors Corporation (GM). Falling between Chevrolet and Oldsmobile on the corporate ladder, Oakland introduced Pontiac in 1926 as a companion model line. The Pontiac line proved to be popular and in 1932, General Motors discontinued Oakland, and the Pontiac Motor Car Company was officially formed.

Pontiac Motor Car Company became Pontiac Motor Division (PMD) in the early 1930s and developed a 268-ci "Straight-8", which was quite reliable.

V-8 Development

In 1954, General Motors leaned heavily on Chevrolet and Pontiac to develop V-8 engines for the 1955 model year and provided them with all available resources to ensure success.

Pontiac settled on a 3.75-inch bore and 3.25-inch stroke to produce a 287-ci package. Pontiac management urged that the design allow for future displacement increases, which eventually permitted the basic combination to displace 455-ci in a matter of years.

Oakland introduced the Pontiac in 1926 to complement itself. Pontiacs sold so well that the Oakland line was canceled within a few years. This particular 1926 Pontiac is part of the General Motors Heritage Collection and was the first Pontiac ever produced. It's the patriarch of the Pontiac hobby that maintains a strong following today.

The V-8 Reaches Production

Pontiac's 287 block featured five 2.5-inch-diameter main journals that support a forged-steel crankshaft. The left-hand cylinder bank was offset rearward. This allowed the distributor to be

The new OHV V-8 "Strato Streak" engine debuted in 1955. Displacing 287 ci, the entry-level 2-barrel mill was packed full of cutting-edge features and was rated at 173 hp. An optional 4-barrel carburetor was made available midyear, which increased horsepower to 200.

mounted on the right side of the block, exposing its driven gear to upward thrust, eliminating the need for a machined thrust surface within the block.

The cylinder heads incorporated 1.78-inch intake and 1.5-inch-diameter exhaust valves. A reverse-flow cooling system directed coolant toward the cylinder heads before the lower end of the engine in an attempt to extend exhaust valve life.

The block's lifter galley was fed with a relatively high volume of oil to accommodate hydraulic valve lifters to provide quiet, consistent, and maintenance-free valvetrain operation. The rocker system, consisting of a stamped-steel rocker arm pivoting on a single stud, was one that many other manufacturers adopted for their V-8s in later years.

Engines featuring a compression ratio of 7.4:1 were rated at 173 hp. A machined (decked) cylinder head was utilized to boost compression ratio to 8:1, and increased horsepower to 180. A 200-hp version was introduced in mid 1955 when a 4-barrel carburetor was made available.

Growth Spurts

Engine displacement grew to 316.6 for 1956, and horsepower increased to 205 for the 2-barrel and 227 for the 4-barrel. Optional dual exhaust added about 10 hp to either engine. Pontiac created its first extra-horsepower package in mid 1956. With more compression, dual 4-barrel carburetors, and a special camshaft, the combination boosted horsepower to 285.

Semon "Bunkie" Knudsen was appointed as Pontiac's general manager toward the end of the 1956 model year. With the promotion came instructions to drastically change the division's image beginning with 1957. In addition to revisions he made to body styling, Knudsen hired Pete Estes—who was part of Oldsmobile's very successful high-performance V-8 program—as chief engi-

The 1956 285-hp Strato-Streak Engine

The 285-hp Strato-Streak proved to be very successful when raced as intended. At Bonneville, Ab Jenkins averaged more than 126 mph for 100 miles, and 118.34 mph over the course of 24 hours in his 1956 Pontiac. And one report claims a top speed in excess of 132 mph was recorded at Daytona Beach during the Flying Mile competition.

At the request of customers, Pontiac introduced its first "extra-horsepower" combination in February 1956. The basic 316.6 block featured specific cylinder heads, a new intake manifold, two 4-barrel carburetors, a high-flow air cleaner, and a specific camshaft and valvetrain components for high RPM. The engine was rated at an astounding 285 hp—a power increase of more than 40 percent over the previous year. Hobbyists intimately familiar with the package claim that approximately 200 were installed. Dealer paperwork for a particular vehicle may be the only way to document originality. (Photo Courtesy Don Keefe)

Bunkie Knudsen was brought in to change Pontiac's image. Though his influence is felt in 1957 model year styling, this 1958 Bonneville represents the direction Pontiac was headed during the late 1950s and into the 1960s. The low-slung body with flashy accents made the Pontiac seem as if it's moving forward while standing still.

neer, and a young John Z. DeLorean as assistant chief engineer.

Even though General Motors agreed to adhere to the Automobile Manufacturer's Association (AMA) 1957 recommendation, which stated that manufacturers should refrain from encouraging any type of speed or acceleration-related con-

Tri-Power was introduced in 1957 to give Pontiac a youthful appeal to young hot rodders. Boasting excellent street manners and economy while operating on its center 2-barrel carburetor, and strong acceleration as its end 2-barrels progressively opened, it was well marketed and quite popular with buyers through 1966. It is truly a Pontiac trademark.

tests, Pontiac increased V-8 displacement to 347. A trio of 2-barrel carburetors, heralded as "Tri-Power," was made optional, increasing horsepower to 290. The top performer designated for NASCAR-type competition still utilized dual 4-barrel carburetors and cranked out 317 hp.

The Super Duty Era

Displacement increased to 370 for 1958 and horsepower increased accordingly. The engine grew yet again in 1959 to 389 ci, and it remained at this size for 1960. That year also initiated Pontiac's illustrious "Super Duty" era.

A complete over-the-counter parts package with the sole intent of producing a competition-only 389 was released in 1960. It employed the same four-bolt block used in Tri-Power applications, but the Super Duty package filled it with a forged-steel crankshaft and connecting rods. High-flow cylinder heads produced 10.5:1 compression and a solid-lifter camshaft with 1.65:1-ratio rocker arms controlled valve action. High-flow cast-iron exhaust manifolds (with separate collectors that could be uncapped) were offered, and a single 4-barrel or Tri-Power intake manifold was also used.

Actual output of the Super Duty 389 wasn't published, but certain racing series required that manufacturers provide a horsepower rating, so Pontiac arbitrarily rated the 4-barrel engine at 348 hp while the Tri-Power was rated at 363. The combination was quite successful as Pontiacs went on to win 7 of the 44 NASCAR races in the 1960 season, and Pontiac ad-man Jim Wangers piloted a 1960 Ventura to the NHRA National championship.

The 389 was carried over into the 1961 model year and the Super Duty package went on. Intake manifold choices were limited to a single 4-barrel or Tri-Power. While both were previously cast iron, for 1961 they were cast in lighter weight aluminum. Either engine was rated at 368 hp.

The Super Duty combination proved to be lethal. Pontiac was dominating drag strips around the country and it captured 30 of 52 NASCAR race wins during the 1961 season. The 421-ci V-8 debuted in Pontiac's parts books toward the end of the 1961 model year and was a direct response to Chevy's 409, Ford's 406, and Mopar's 413. It's generally accepted that none were factory installed.

The Super Duty 421 was comprised of a new 4.09-inch bore block with four-bolt main caps, a forged-steel 4-inch stroke crank, forged-steel connecting rods, and forged-aluminum pistons. It utilized the same cylinder heads as the 389, which pushed compression to 11.0:1, a cast-aluminum dual 4-barrel intake manifold, and a solid-lifter camshaft. Rated at 373 hp, the package made Pontiac an even greater threat on the race track.

If success is measured by competitive wins, Pontiac was the manufacturer to beat in 1962. The possibilities seemed limitless with the persuasive John DeLorean having been promoted to chief engineer the previous year.

Until that point, certain forms of racing simply required that a given component have a factory part number for it to be legal for competition. That translated into a flood of aftermarket components arbitrarily hung with manufacturer part numbers with the sole intent of satisfying these governing bodies. It didn't take long for the National Hot Rod Association (NHRA) to catch on and revise its ruling for certain stock classes. Such components not only needed to have a manufacturer's part number, but they had to be factory-installed too.

Pontiac responded with the announcement of a series of "special purpose" factory-built vehicles. This simply meant that Pontiac installed both the SD-389 and SD-421 into vehicles on its assembly line. The SD-389 was limited to a single 4-barrel while the SD-421 used dual 4-barrels. Both engines received the McKellar number-10 camshaft, which was much like a solid-lifter version of the hydraulic number-041 of later years. The SD-389 was rated at 385 hp, while the SD-421 was at 405.

The 1963 Super Duty 421 received new pistons, which raised the compression a full point over the previous year to 12:1. The single 4-barrel version was rated at 390 hp, while the dual 4-barrel unit was rated at 405 hp as in 1962, even with the increased compression ratio. A separate dual 4-barrel engine with 13:1 compression was also available in 1963, and it was (under)rated at 410 hp.

Initially available through its dealership parts counters for the Tri-Power 389 in 1960, Pontiac's Super Duty package included a forged-steel crankshaft and connecting rods, a solid-lifter camshaft, and high-flow cylinder head and exhaust manifolds. The package was available in 1961 with few enhancements, and it continued as a parts department purchase.

Royal Pontiac in Royal Oak, Michigan, capitalized on Pontiac's success on the drag strip early on. With the assistance of Pontiac ad-man Jim Wangers, Royal Pontiac became intertwined with the Division and had a direct line to its Engineering department. This particular 1961 Ventura, campaigned by Royal Pontiac and driven by Wangers, may be the only 1961 Pontiac to receive a factory-installed Super Duty 389.

CHAPTER 1

To maintain competitiveness on the track with the larger engines produced by other manufacturers, Pontiac released a 421 version of its Super Duty package through its parts departments toward the end of the 1961 model year. To comply with rules imposed by racing associations, both it and the Super Duty 389 became factory-installed options in 1962. Available only with two 4-barrels, the SD-421 was rated at 405 hp and made quick work of the competition.

When General Motors pulled the plug on factory-backed racing, Pontiac moved forward with its street performance program. A new vehicle based on the concept of combining a large-cube engine with a compact body produced the GTO for 1964. Corporate regulations limited engine size to 389 in the Tempest platform. Many purists argue that the GTO was America's first muscle car, initiating the era of serious performance.

In addition to the induction changes for 1967, Pontiac also increased displacement and revised the cylinder head valve angles to improve airflow and further enhance output. The GTO's 360-hp 389 grew to 400 and the new 4-barrel engine was every bit as powerful as the former Tri-Power mill. (Photo Courtesy Tom DeMauro)

Concerned with maximum performance, little regard was given to cold-weather operating characteristics of the Super Duty engine. As such, most Super Duty Pontiacs were purchased with the sole intent of regularly competing on the race track. Campaigned by Gay Pontiac, driver Don Gay claimed the 1963 NHRA National Championship in class A/Stock with this 1962 Catalina.

Pontiac released a revised cylinder head for its Super Duty engines early in 1963. Boasting improved exhaust flow, it had no official effect on output rating.

The Corporate Edict

The bottom seemingly fell out of the performance car market at General Motors in January 1963. GM Chairman Frederic Donner issued a memo to all divisions reaffirming its position on the AMA agreement against racing.

This meant that manufacturers couldn't openly sponsor race programs, and that experimental components had to be covertly supplied to racers. Vehicles were produced for a few months even after the edict was imposed, however, as evident with the "Swiss Cheese" Catalinas, but General Motors was serious and its Divisions were forced to comply. Pontiac utilized the experience gained from the Super Duty program to create its newest performance street engine, the 421 High Output (421 H.O.).

The 421 H.O. was essentially a detuned SD-421. It consisted of a four-bolt block and number-716 cylinder heads, which were similar to the Super Duty units but contained an exhaust crossover to improve cold-weather operation. An aggressive hydraulic-lifter camshaft was employed, and it utilized Tri-Power induction and high-flow cast-iron exhaust manifolds. It was rated at 370 hp.

A new performance Pontiac entered the market in 1964 and it changed the industry forever. The new "GTO" was built on the intermediate Tempest platform. The 421 H.O. seemed to be a natural fit for the new GTO, but GM's power-to-weight-ratio standards limited maximum engine size to just 389, so Pontiac simply created a 348-hp 389 using components familiar to the 421 H.O.

DeLorean Promoted

John DeLorean was promoted from the position of Chief Engineer to General Manager in 1965. His rebellious attitude allowed him to push many of his technological visions through to production.

Pontiac engines had typically been painted a shade of green or blue up to this point. DeLorean wanted buyers to find the vehicle's engine as attractive as its exterior, so in 1966 the Pontiac V-8 was painted a light metallic-blue, and a chrome-plated air cleaner and valve covers were added in performance applications.

Though the same engine packages were carried over, the Tri-Power's end carburetors were made larger for improved airflow, and that boosted horsepower of engines so equipped.

A dealer-installed "Ram Air" package was made available for the GTO late in the 1965 model year. The hood scoop insert was cut open, allowing the engine to ingest cooler outside air, conceivably producing more power. The Ram Air package became a Pontiac trademark for years to follow. It never increased the advertised output rating, but it certainly offered a performance benefit.

Major V-8 Design Changes

Pontiac was forced to abandon its signature Tri-Power induction system in 1967 when General Motors banned multiple carburetion on all vehicles except the Corvette. Rochester's new Quadrajet 4-barrel was Pontiac's choice for its performance applications.

To be sure the new 4-barrel engines performed at least as well as the previous 360-hp Tri-Power engine, airflow was improved by reducing piston-to-valve angle from 20 to 14 degrees, and increasing valve diameters from 1.92/1.66 inches to 2.11/1.77 (intake/exhaust, respectively) in performance applications. Streamlined exhaust manifolds were used to improve flow, and the block bore diameter was increased .030 inch, which boosted displacement to 400 ci.

In addition to those changes, Pontiac completely redesigned the intake manifold using the 1960s Super Duty 4-barrel manifold as a template. The dual-plane design featured long, smoothly contoured runners to produce maximum torque at low speed. Though there was large push to cast the manifold in aluminum to save weight, cast iron was ultimately used to

General Motors banned the use of multiple carburetion in 1967. In an attempt to maintain performance, Pontiac developed a new intake manifold and specified the new Rochester Quadrajet carburetor for its performance applications. Capable of flowing 750 cfm, it was an efficient design that was used well into the mid 1980s. (Photo Courtesy Tom DeMauro)

Tunnel Port Ram Air V

At the urging of John DeLorean, Pontiac developed a high-RPM, maximum performance engine package during the late 1960s, featuring tunnel-port cylinder heads using Ford's design as a template. Dubbed Ram Air V (R/A V), the package included a reinforced block, a forged-steel crankshaft and connecting rods, and cylinder heads with huge intake ports with round openings and individual exhaust ports.

Several variations were tested, and displacement ranged from 303 to 428 depending upon the combination tested. A single 4-barrel or dual 4-barrel Holley carburetor(s) was mounted atop a cast-aluminum intake manifold. One engineering report suggests that peak horsepower eclipsed 565 from a single 4-barrel 400!

The R/A V 400 was intended to be a regular production option in certain 1969 models, and enough components were produced to build a small number of complete engines in anticipation of customer orders. The R/A V project was scrapped just before the engine package was released for vehicle assembly, and several complete engines and a number of individual components were sold through dealership parts departments until supply was exhausted.

Though rare and quite valuable, it isn't completely uncommon to occasionally find a vintage Pontiac at race tracks or show fields with individual R/A V components or even a complete R/A V engine. In fact, at least two companies are engineering or producing cast-aluminum R/A V cylinder heads with modern enhancements (see Chapter 6).

The Ram Air V (R/A V) was Pontiac's tunnel-port effort. Developed in the late 1960s, the 400-ci variant reached production and nearly reached vehicle assembly plants before the project was canceled. Its demise was most likely related to emissions standards, and a small number of complete engines and several individual components were made through Pontiac Parts Department. (Photo Courtesy Tom DeMauro)

quell reliability concerns and maximize cold-weather operating characteristics.

The 421 was also slightly affected for 1967. As the full-size offerings grew in size, they required even more horsepower to maintain performance. Pontiac increased the 421's bore to 4.12 inches, which produced 426.5 ci when combined with the 4-inch-stroke crank. To maintain its own identity in a market filled with the 426 Hemi and 427 Chevy, Pontiac's was billed as a "428."

The Round-Port Era

The Ram Air 400 available in the GTO and new Firebird was very much the same for 1968, and it came to be known as "Ram Air I" when a new Ram Air engine was brought to market in May 1968. The Ram Air II (R/A II) was rushed through the development process so Pontiac could give its customers a high-winding V-8. It borrowed technology from a new high-performance engine that Pontiac was developing for 1969, which contained some very unique pieces aimed at reaching its intended 6,000-rpm limit.

The R/A II featured all-new cylinder heads with redesigned round exhaust ports. The port work improved exhaust air-flow by about 10 percent over a comparable D-port, and the outlet shape was intended to make fitting tubular headers easier for racers. The valvetrain was comprised of specific heavy-duty components, and the new number-041 hydraulic-lifter cam was teamed with 1.5:1 rockers to produce .470 inch of valve lift. The combination was rated at 340 hp for the Firebird and 366 for the GTO.

Two new performance engines were introduced for the 1969 Firebird and GTO, and both carried over into 1970 with minimal changes. Rated at 335 hp for the Firebird and 366 in the GTO, the 400 H.O., or Ram Air III (R/A III) as it was later known, utilized D-port cylinder heads and high-flow exhaust manifolds.

Pontiac's top engine option in 1969 and 1970 was its Ram Air IV (R/A IV), which had a solid 6,000-rpm operating limit. The intake-port roof of the round-port R/A IV cylinder head was raised 1/8 inch and the intake port volume was increased from 153 to 180 cc, which allowed it to operate at its intended limit.

The high-flow R/A IV cylinder heads were complemented by a new cast-aluminum 4-barrel intake manifold with enlarged runners and separate cast-iron heat crossover. The 041 camshaft teamed with 1.65:1 ratio rocker arms produced a gross valve lift of .520 inch—the most ever used in any production Pontiac engine. The mill was rated at 345 hp for the Firebird and 370 hp for the GTO.

Increased displacement was required to motivate GM's full-size cars, which continued getting larger year after year. Pontiac's 428 grew to 455 in 1970 by increasing bore size .030 inch and replacing the 4-inch-stroke crank with a 4.21-inch unit. The 455 H.O. was comparable to the previous year's 428 H.O. and availability was limited to larger models and the GTO. The R/A III and R/A IV continued as Pontiac's top engine options in the GTO and Firebird model lines.

Living with Low Compression

To regulate emissions, General Motors imposed a compression-ratio

The Ram Air II evolved into the Ram Air IV for 1969. New cylinder heads with larger intake ports, cast-aluminum intake manifold, and 1.65 rocker arms for the 041 cam were all intended to further improve the 400's performance. Available in the GTO and Firebird, the 370-hp engine with nearly 10.5:1 compression was a stout performer. Such examples are highly coveted by collectors today.

The second-generation Firebird was introduced in 1970 and the Firebird Trans Am quickly became Pontiac's premier performance model. With desirable rarity from a collector's perspective, and capable of stout performance and excellent handling characteristics right off the showroom floor, the entire model line remains extremely popular today.

HISTORY OF THE PONTIAC V-8 1955–1981

To combat certain forms of tailpipe emissions and to keep insurance premiums in check, General Motors imposed a corporate compression ratio limit of 8.5:1 for 1971. Pontiac increased the displacement of its Ram Air IV to offset the loss of performance associated with lesser compression to produce that year's 455 H.O. Rated at 335 hp, it used many Ram Air IV–type components and included revised round-port cylinder heads and an 068 camshaft.

An exhaust gas recirculation (EGR) valve is a vacuum-operated emissions control device located on the intake manifold of every Pontiac V-8 from 1973 forward. Its purpose is to allow metered amounts of exhaust gas to reenter the cylinders during certain operating conditions, limiting the formation of a specific pollutant. By design it's nonfunctional at full throttle, so it shouldn't have any effect on performance.

Even when forced to comply with emissions regulations, Pontiac shocked the industry when it released the Super Duty 455 in 1973. Featuring such components as a specially reinforced block, forged pistons and connecting rods, and new high-flow cylinder heads, the round-port engine was capable of running at 6,000 rpm. Availability was limited to the Firebird Formula and Trans Am in 1973 and 1974. Fewer than 1,300 were built during its two-year run.

cap of 8.5:1 for the 1971 model year. That signaled the end for the R/A IV. Pontiac knew that increasing displacement meant similar horsepower could be attained at a lower RPM. By combining R/A IV–type components with the 455, Pontiac created the new 455 H.O.

Adhering to the imposed compression ratio limit, the 455 H.O. featured modified round-port R/A IV cylinder heads, a cast-aluminum intake manifold, and high-flow exhaust manifolds. The 068 camshaft was chosen to maximize low-end torque, and specific hydraulic lifters were used to effectively limit engine speed to no more than about 5,500 rpm, quelling warranty claims from overextended operation. The new engine carried a gross rating of 335 hp, and 305 hp at the new "net" rating, which more closely represented engine output when installed in a car. It remained Pontiac's top engine option for 1972.

In response to more stringent exhaust emission standards, exhaust gas recirculation (EGR) was introduced for 1973. It consisted of a valve mounted on the intake manifold that allowed metered amounts of inert exhaust gas to re-enter the combustion chamber. The biggest news that year, however, was Pontiac's newest performance engine, the Super Duty 455.

The Second Super Duty Era

The SD-455 was a max-performance effort designed to operate at 6,000 rpm. Additional material was added to the SD-455 block to increase overall rigidity, and it contained a provision for dry sump oiling at the rear. A nodular iron crankshaft with deep-rolled fillets was employed. It was retained by four-bolt main bearing caps. Specific forged-aluminum pistons were complemented by beautiful forged-steel connecting rods.

The SD-455 cylinder heads' intake-port volume was increased to 186 cc to allow for 6,000-rpm operation, and the exhaust ports were precisely modified to maximize flow. The intake port was so wide near the entrance that its sidewall actually broke into the adjacent pushrod guide passage, and a thin-wall steel sleeve was pressed in to seal it. Specific valvesprings and high-quality 2.11/1.77-inch valves were also used.

A new 800-cfm Quadrajet and a specific cast-iron intake manifold with enlarged runners were used with the

CHAPTER 1

To improve the performance of its 301, Pontiac installed a turbocharger for the 1980 model year. It added more than 50 hp to the naturally aspirated 4-barrel mill, taking the total to around 200 hp. A number of special components were used to accommodate the added cylinder pressure that occurs under boost conditions. Availability was limited to the Firebird Formula and Trans Am.

SD-455. A 041-spec hydraulic camshaft was used throughout development and testing, lending to its 310-hp rating, but when it finally reached production in May 1973, a 744-spec cam was used to ward off emissions concerns, and the engine was subsequently rerated at 290 hp. Availability was limited to the Firebird model line. Only 295 SD-powered Firebirds were produced in 1973, and all were the 290 hp variety. An additional 1,000 Super Duty Firebirds were produced in 1974.

Emissions and Economy

New federal emission standards shook the industry during the 1975 model year. Pontiac utilized a single exhaust catalyst and a compression ratio of just 7.6:1 to ensure total compliance. High-ratio rear axle gearing was used to keep engine speed relatively low, which lessened emissions and improved fuel economy. Performance suffered and the 455 was emasculated to just 200 hp.

A change that affected all 350 and 400 engines was the introduction of the "lighter" block castings in mid 1975. Pontiac knew that high-revving engines had become a distant memory, so in an attempt to shed overall vehicle weight, material was removed from low stress areas of the block. The blocks are reliable for normal duty applications, but should not be used in any high-performance effort.

Another significant change occurred in mid 1976. Pontiac eliminated the common harmonic balancer on most 350 and 400-ci engines backed by an automatic transmission. A crankshaft hub was used in its place and it served as an accessory drive and contained a top dead center (TDC) timing mark.

Emissions-Compliant Performance

The 455 was discontinued after 1976 and a high-performance 400 took its place in 1977. Rated at 200 hp, the new T/A 6.6 received a compression boost to 8:1 using 350-spec 6X cylinder heads, a unique camshaft, and specific carburetor and distributor settings.

With heavy emphasis placed on maximizing fuel economy, Pontiac developed a small-cube V-8 in a lightweight package to complement the downsized models it would introduce in the near future. The short-deck 301 was Pontiac's answer to Chevy's 305 and engineers had no intentions of it being a performance mill. The svelte block was filled with a crankshaft that had only one large counterweight at each end and cast connecting rods.

Revised camshaft timing and exhaust enhancements bumped the 400-ci T/A 6.6 to 220 hp for 1978, and durability issues arose with the lighter-weight 400 block. Pontiac revived a former 400 block casting specifically for 1978–1979 T/A 6.6 engines. These castings were as good as earlier units and can be identified by the "XX" cast into several locations on the block. Pontiac's 350 was discontinued after the 1977 model year, and the last 400 blocks were cast on Thanksgiving weekend 1977 and stockpiled for use for the remainder of 1978. Several thousand were set aside for the 1979 Trans Am.

The End of an Era

The 301 was Pontiac's only V-8 left in production by 1979. The T/A 6.6 was available for most of the model year, but once the supply of stockpiled 400s was exhausted, it signaled the end of the big Pontiac V-8.

A turbocharger was added to the 301 to give the 1980 Trans Am an injection of performance. With turbo boost limited to less than 10 pounds, the 301 was rated at around 200 hp in both 1980 and 1981. General Motors ceased Pontiac V-8 production in March 1981. It was the final chapter in a saga that started in 1955 and concluded after 14,624,886 engines were produced.

Pontiac Racing

Pontiac always maintained a performance image and a great number of hobbyists competed regularly with their Pontiacs in various stock and modified classes.

Some of those who have successfully campaigned Pontiacs during the 1950s, 1960s, and 1970s include Arnie Beswick, Truman Fields, Jim Hand, John Angeles and Pete McCarthy, Art Peterson, Nunzi Romano, Milt Schornack, Mickey

HISTORY OF THE PONTIAC V-8 1955–1981

Arnie Beswick drove a number of different Pontiacs during the 1960s and 1970s. Competing in classes ranging from Stock to Factory Experimental or others, Beswick capitalized on his Midwest roots and dressed like a farmer in an attempt to fool the competition, earning his trademark "Farmer" nickname. He remains quite active and often attends various major Pontiac show and/or race events across the country. He almost always has one of his Pontiacs with him.

Mickey Thompson was so serious about Pontiac performance during the early 1960s that he took it upon himself to develop and produce a specific cylinder head with a hemispherically shaped combustion chamber, which was sold by his Long Beach, California–based Mickey Thompson Equipment Company. A complete line of accessories was also available, which included pistons, valve covers, and intake manifolds. This particular engine is owned by hobbyist Jack Gifford.

According to vintage Micky Thompson (M/T) sales literature, the Pontiac hemi head allowed the use of much larger valves to improve airflow and placed the spark plug near the center of the combustion chamber to promote a more consistent burn. Thompson claimed an increase of more than 100 hp was possible when compared to modified Pontiac heads.

Kauffman Racing Equipment produces a wide variety of heavy-duty components for those wishing to compete at the top levels. This 535-inch engine features an aluminum block, canted-valve cylinder heads, and dual carburetors. It generates more than 1,100 hp and has run the quarter-mile in as quick as 7.20 seconds at more than 190 mph in a 2,200-pound race car with a Pontiac GXP body.

Butler Performance has gained a reputation as a premier Pontiac engine builder providing hobbyists with potent combinations ranging from dedicated street engines to max-performance boosted V-8s. Rodney Butler's 482-ci Pontiac features twin turbochargers that provide 40 pounds of boost. The 2,880-hp beast has propelled his 2,700-pound 1963 LeMans to a quarter-mile best of 6.27 at 228 mph. (Photo Courtesy Don Keefe)

Thompson, Jess Tyree, Jim Wangers, and Arlen Vanke. These guys drove some of the quickest Pontiacs to ever make a pass down the drag strip in their day.

Several companies specialized in improving Pontiac performance, including Baldwin-Motion, H-O Racing, Leader Automotive, Nunzi's Automotive, and Royal Pontiac. These companies, and a few others, can be credited with keeping Pontiac V-8 performance flame alive during the smog-era and the years immediately following its discontinuance. They paved the way for today's aftermarket companies, which produce the components that give new meaning to the term max-performance Pontiac V-8.

CHAPTER 2

BLOCKS

The stock Pontiac V-8 block is constructed of cast iron. When introduced in 1955, it displaced only 287 ci. Its robust design allowed immediate adaptability to future displacement increases. Though external dimensions remained very close to the original 287 design, the V-8 eventually displaced as much as 455 ci by increasing bore diameter and stroke length. With today's long-stroke aftermarket crankshafts, a total displacement near 500 ci using a stock block is quite possible.

Block Design

The basic block is compact and rigid. Its deck height is relatively tall at 10.24 inches as measured from the crankshaft centerline upward. The deck surface is very thick and uses ten 1/2-inch-diameter bolts, which pass through the deck surface and into individual bosses within the coolant jacket walls to retain the cylinder heads. The individual bosses limit cylinder distortion when the head bolts are torqued appropriately. The holes are also blind and do not extend into the water jacket, which would otherwise require thread sealer during installation.

A symmetric 90-degree design allows for easy and precise block machining. The main journal saddles are quite thick, which improves block rigidity and allows the use of long connecting rods for good rod-to-stroke ratio. From 1959 on, engines displacing between 326 and 400 inches contained a main journal diameter of 3 inches,

The stock Pontiac V-8 block evolved from its original displacement of 287 ci in 1955 to as much as 455 by 1970 without grossly deviating from its original external dimensions. Its robust main saddles and thick deck surface are among the many features designed to improve rigidity and durability. Original 400 and 455 blocks are the most popular for modern performance builds up to 600 hp or slightly more.

while 421-, 428-, and 455-ci engines featured 3.25-inch journals. The main bearing caps are fastened to the block by two large bolts in most instances and located by dowel pins to prevent the caps from wandering during high-speed operation.

BLOCKS

Block Specifications

Pontiac used a variety of bore and stroke combinations to achieve specific amounts of displacement for its V-8 over the years. Main bearing diameter varied from 2.5 inches to as much as 3.25 depending upon the application. This chart provides the original dimensions of Pontiac's regular production V-8s.

Year	Displacement (ci)	Bore (inches)	Stroke (inches)	Main Journal Diameter (inches)
1955	287	3.75	3.25	2.50
1956	316	3.94	3.25	2.50
1957	347	3.94	3.56	2.62
1958	370	4.06	3.56	2.62
1959–1966	389	4.06	3.75	3.00
1963–1966	421	4.09	4.00	3.25
1963–1967	326	3.72	3.75	3.00
1967–1969	428	4.12	4.00	3.25
1967–1979	400	4.12	3.75	3.00
1968–1977	350	3.88	3.75	3.00
1970–1976	455	4.15	4.21	3.25
1977–1981	301	4.00	3.00	3.00
1980–1981	265	3.75	3.00	3.00

Most Pontiac blocks from 1970 forward have four or five engine-mount bolt holes on each side, which allows easy installation into any chassis. Earlier blocks, such as this 1968 400, and even some very late 1970s blocks have a limited number, and that can present installation issues. While adapter kits are available, they may compromise durability in extreme applications. Be sure the block you select has the correct mounting points for your chassis.

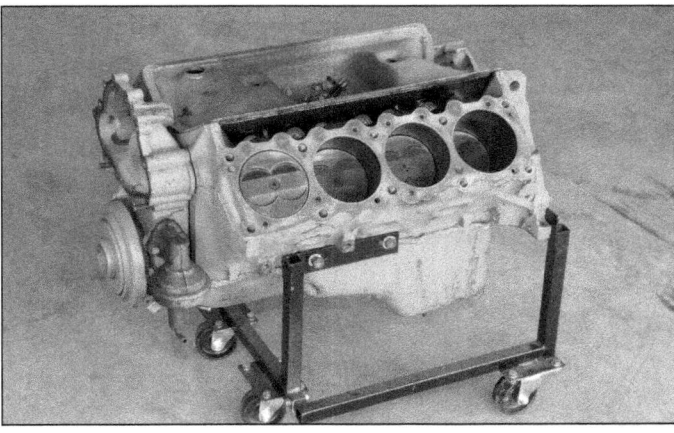

When selecting a block for your build, a complete running engine makes a suitable candidate. I found this 1974 400 locally for $250 and confirmed that it retains its original 4.12-ci bore after removing the cylinder heads. With some hunting, you can likely find a similar bargain. I don't recommend spending more than $500 to $700 for any Pontiac engine unless for a numbers-correct example.

Oil is collected in a sump at the rear of the oil pan. It's drawn in, pressurized, and dispersed by a positive-displacement rotary pump that's driven by the camshaft. The oil is filtered before it passes across the rear of the block and into a machined galley that runs parallel to the crankshaft on the left side. As the oil travels forward, it lubricates the camshaft, crankshaft, and left-side lifter bores. It then crosses over to the right side of the block at the front journal before traveling toward the rear lubricating the remaining lifter bores.

The original V-8 block design saw a number of changes over the years, and that can limit direct component interchange. That includes a variety of displacements, revised engine mounting points, transmission bellhousing bolt patterns, starter locations, main journal diameters, and cooling system variances. Other differences less concerning include the number of water jacket plugs and rocker arm oiling path.

Selecting a Stock Block

When considering a block for your particular build, any casting from 1965 and later is probably the best choice. Blocks of this era are constructed of

Beginning in 1968, Pontiac located the block casting number on a ledge just rearward of the number-8 cylinder. Many numbers were available. This particular block, number 488988, was used in late-1974 400 applications and contains all the desirable characteristics for performance use.

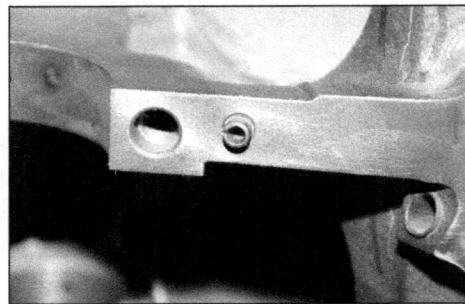

A block I recommend avoiding is number 500557, which was introduced in 1975 and commonly used through 1978, at which point 400 production ceased. Thickness in certain areas was reduced to save weight and roll pins were used in place of dowels for main cap retention. While completely suitable for low-RPM production applications and builds up to 400 hp or so, it doesn't possess the durability required for high-performance use.

Common Block Casting Numbers

If you're considering a stock Pontiac block for your build, identify the casting you're starting with. This listing contains the original block number for many of Pontiac's regular production castings. It's found on the ledge near the distributor hole through 1967 and behind the number-8 cylinder from 1968 on.

Year	Block Casting	Displacement (ci)
1964	9773153	326
1964	9773155	389
1964	9773157	421
1966	9782611	421
1967	9786135	428
1967	9786133	400
1967	9786339	326
1970	9799140	455
1970	9799914	400
1970	9799916	350
1971	483677	455 H.O.
1965–1966	9778791	421
1965–1966	9778789	389
1965–1966	9778840	326
1968–1969	9792968	428
1968–1969	9792506	Ram Air 400
1968–1969	9790071	400
1968–1969	9790079	350
1971–1972	481990	350
1971–1974	485428	455
1971–1975	481988	400
1973–1974	490132	SD-455
1973–1974	488986	350
1975–1976	500813	455
1975–1977	500810	350
1975–1978	500557	400
1977–1978	568557	400

high-quality iron and contain all of the desirable characteristics to accept most modern aftermarket Pontiac components and accommodate easy chassis installation. Some earlier blocks may be of similar or better quality (particularly the 1960s Super Duty castings) but such examples are rare and may present compatibility concerns with today's parts.

Though 400 and 455 blocks are most popular for modern performance builds, the 455 was far more desirable for many years for its larger displacement and cost reflected it. It wasn't uncommon to pay several hundred dollars for a usable core with standard dimensions. While the 455 allowed Pontiac racers to remain competitive with larger-cube engines of other makes, reliability issues in race applications were common. The area required to accommodate 3.25-inch-diameter main journals created excessive friction and was difficult to properly lubricate at high speed. The wider journal also compromised main saddle rigidity.

The somewhat-recent introduction of long-stroke aftermarket crankshafts with 3-inch main journals has revolutionized the hobby. When compared to a 455, a 400 block is far more plentiful,

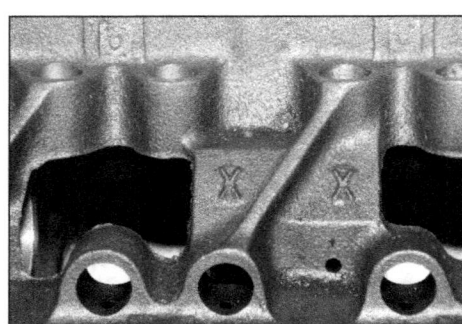

To improve the durability of its 1978–1979 T/A 6.6 engines, Pontiac revived the 481988 casting, which was beefier than the 500557 used elsewhere. To identify the new casting, Pontiac added a large "XX" in several places. That includes on the lifter bore, which otherwise commonly contained the last two digits of engine displacement (such as "55" denoting 455). The XX-481988 blocks are as durable as any earlier block, and can be considered an excellent choice for performance use.

making it cheaper and easier to find. Any performance enthusiast or competitive racer can easily achieve displacement greater than 460 ci using a typical 400 block and an aftermarket crankshaft. The smaller journals reduce bearing speed and improve block integrity, giving racers the performance and reliability required for high-speed operation.

The deck surface of a Pontiac block is very thick, and can measure more than .375 inch. Surfacing to remove irregularities and improve gasket seal, which is known as "decking," should be part of any rebuild. Slightly more can be removed so the deck surface and piston face are level to maximize quench area and promote greatest performance. Gross decking can affect compression distance on subsequent rebuilds, however.

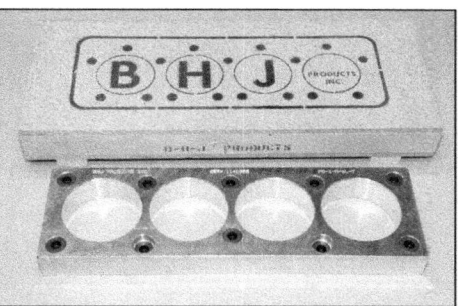
A honing plate (or torque plate) is an excellent investment if you're serious about performance. It replicates the load a block sees when a cylinder head is installed. A quality machine shop uses it to account for cylinder distortion during machining, but it can be beneficial for home use too. That includes curing time for water jacket filler and for light cylinder honing during a refresh. BHJ offers steel and alloy units for the Pontiac V-8.

All cylinder machining should be performed with a honing plate installed. Most Pontiac blocks tolerate an overbore of at least .030 inch, and possibly up to .060 inch or slightly more. Since production variances and core shift are normal, I recommend measuring the cylinder wall thickness using an ultrasonic tester before machining. That allows you to calculate thrust wall thickness and ensure it's at least .120 inch after boring and honing.

While most any 400 block produced between 1967 and 1978 is a suitable candidate for high-performance use, there's one block casting I strongly suggest avoiding entirely. To shed total vehicle weight, Pontiac revised its 400 during the 1975 model year. The svelte 400 block is easiest identified by its 500557 casting number, and lacks the durability required for high-performance applications.

When searching for a stock block, purchasing a complete and running engine is fine so long as you're aware that you have no way of predetermining the block's condition and capability. It may be easiest to find a casting that's completely disassembled and thoroughly cleaned so every portion of it can be thoroughly inspected. A grungy block must be properly cleaned so the block can be inspected to determine if it's a suitable candidate.

Once perfectly clean, the block should be magnetically checked for cracks, and then closely inspect for porosity or rust issues externally as well as in the water jackets. A bright light and small mirror can provide you with an idea of what the block looks like internally, particularly toward the bottom of the cylinders. You should also inspect the main saddles and bulkheads to ensure uniformity throughout and look for casting voids or signs of core shifting.

If you have access to an ultrasonic thickness tester, it's wise to measure cylinder walls thickness on all sides, but the thrust side is most critical. A nominal wall thickness of at least .150 inch on the thrust side, and .100 inch elsewhere should provide plenty of wall material for boring. There's no concern using a block with greater wall thickness, but anything less than about .120 inch on the thrust side after machining may fail in extreme applications.

If you have any doubt about your own opinion, take along an experienced friend or a trusted machinist for a second opinion. It can never hurt!

Stock Block Modifications

A stock Pontiac block is completely adequate for most high-performance builds. Good rebuilding and reconditioning techniques, which includes thoroughly cleaning the oiling passages and precise machining with modern equipment generally produces a block that's well equipped for applications producing up to 600 hp, or slightly more.

The deck should be machined to provide a smooth and consistent surface for optimal gasket seal. Generally measuring somewhere near 3/8 inch thick, it can be machined liberally without worrying about compromising its integrity. Any time material is removed from the deck surface, you need to physically verify that the cylinder head bolts do not bottom out with the cylinder head installed.

Most Pontiac blocks can tolerate an overbore up to .060 inch, and possibly more after proper thickness testing. Boring and honing should always include a honing plate. Also called a "torque plate," it's a rigid piece of steel or aluminum that bolts to the block's deck surface and replicates the load it sees when a cylinder head is installed.

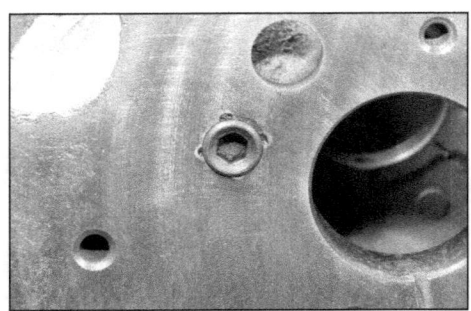

The front oil galley plugs in most stock Pontiac blocks are a press-fit design that are driven into place and staked for maximum retention. Loose-fitting plugs sometimes push out when exposed to very high oil pressure, and that can result in a massive internal oil leak. A popular modification to is tap the front holes to accept common 3/8-NPT pipe plugs. Use caution when installing the plugs, however. If the plugs are threaded in too far, oil flow through the block may be restricted.

Extreme camshaft duration and/or valvespring pressure creates an excessive amount of side loading on the lifter during normal operation. That can cause the lifter bores of a typical Pontiac V-8 block to crack and/or break. The Mega Brace by SD Performance is a bolt-in kit that strengthens the lifter galley in a stock block. It's an excellent design that's used by many professional Pontiac engine builders.

Hard Blok is one of the best water jacket fillers available today. It doesn't generate an excessive amount of heat during the curing process, and doesn't expand when exposed to temperature increases during normal operation. Even though distortion is minimal, I recommend that it be poured and cured before any block machining.

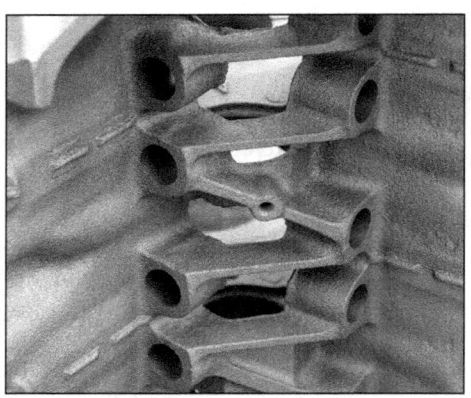

When Pontiac developed its tunnel-port Ram Air V, additional ribbing was added to connect the lifter bore banks in the block's lifter galley, improving rigidity and durability. Though the R/A V never reached regular production, the feature was incorporated into the 1973–1974 SD-455. It's beneficial when running a camshaft with very aggressive specifications.

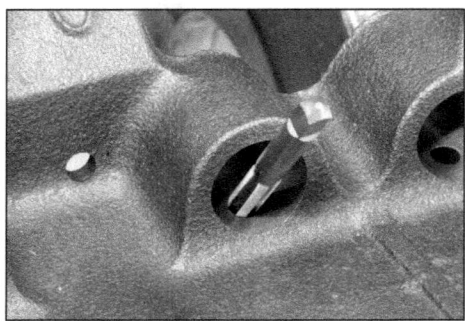

The stock Pontiac block delivers a large volume of pressurized oil for proper hydraulic lifter function. Solid lifters do not require it, and that can require the installation of lifter-bore restrictors that limit oil flow. The modification consists of cutting several threads into the lifter bore using a 1/4-20 or 1/4-28 tap, and then inserting stainless-steel set screws with a .030- to .060-inch hole to effectively limit oil flow. A few manufacturers now offer Pontiac-specific solid lifters that are internally restricted, lessening the need for this modification.

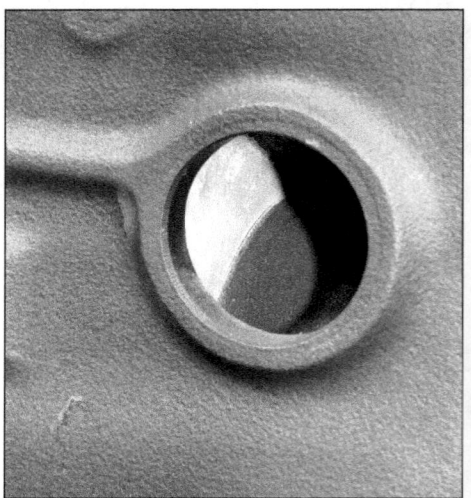

Adding water jacket filler is an excellent way to improve the rigidity of any block. Poured in as a liquid, it encapsulates the cylinder bottoms by replacing the coolant that otherwise surrounds them. Once cured it essentially becomes a permanent part of the block, preventing distortion and absorbing harmonics. Installation of a honing plate and the main caps during the pour is strongly recommended to maintain ideal block dimension during the curing process.

"Hot honing" consists of heating the block until it's near its normal operating temperature (around 200 degrees F) before machining. While it's not practical or necessary for every performance level, hot-honing may benefit those chasing every last horsepower when racing at the most competitive levels.

An aggressive roller camshaft and the valvespring pressure it requires greatly stresses the lifter bores of a stock Pontiac block. Lifter bore brace kits, such as the Mega Brace from SD Performance contain formed plates that are bolted, epoxied, and/or welded into the lifter valley to improve lifter bore strength.

Another common modification performed to the lifter bores in a stock Pontiac block is to tap the lubrication holes and insert small restrictors whenever a solid-lifter cam is used. Stock lifter bores supply a relatively large volume of oil for proper hydraulic lifter operation. Solid lifters simply pass oil through to the rocker arms, and if restrictors are not installed to limit oil flow, oil will freely flow to the top of the engine, reducing the supply available for the main journals. Many Pontiac builders still offer lifter bore restrictor kits, but many modern solid lifters (flat-tappet and roller alike) are restricted internally to control the amount of oil that reaches the top end.

Water Jacket Filler

A stock Pontiac block can be dimensionally unstable if subjected to excessive cylinder wall side loading from a long stroke crankshaft, extreme cylinder pressure, and/or excessive RPM associated with max-performance engines. Another instance may be if the cylinder walls are at the minimum recommended thickness after boring. That instability can lead to a number of operating issues that can negatively impact performance or result in engine failure.

Water jacket filler is a thick liquid that's poured into the coolant jacket of a block, encapsulating the cylinders. Once permanently set, it makes a block less susceptible to flexing, which improves cylinder-wall strength and seal, dampens harmonics, and reduces main-cap walking. If a torque plate is available, installing it just after pouring ensures that the cylinders are located properly during the curing process. Most water jacket fillers expand slightly while curing, and because of that I highly recommend that machining be performed only after the filler has cured completely.

Most builders suggest filling the block to the bottom of large freeze plug holes in the side for maximum durability in applications producing at least 600 hp. It reduces volume of coolant circulating in the block, and that can possibly increase oil temperature slightly. A "short fill" is best suited for street/strip engines producing more than 550 hp where some additional rigidity is beneficial, but still allows for plenty of coolant circulation. It typically consists of filling the block until the filler reaches about

What Are "Splayed" Main Caps?

The outer holes of a typical Pontiac four-bolt block extend into a rather weak area of the main saddle, and that can compromise block rigidity. While it's of little concern when producing less than about 700 hp, it can certainly be an issue in max-performance applications exposed to excessive cylinder pressure and/or high engine speed (RPM), or where severe detonation related to improper tune occurs. If the entire bulkhead isn't rigid enough for the application, cracks can propagate from any hole drilled into the main saddle.

A "splayed" main cap features a modified four-bolt design where the outer retaining bolts are drilled on angle away from the crankshaft centerline and into a beefier area of the block. It allows bolt torque to pull the block from different angles, drawing the bottom end together to improve strength. Splayed caps are generally installed on the center three main journals and Pro-Gram Engineering's billet-steel units are an excellent choice when upgrading a stock Pontiac block. Aftermarket Pontiac blocks use splayed main caps as standard equipment.

When selecting a stock Pontiac block in which to add splayed main caps, it's best to begin with one that's drilled for two-bolt caps only. The lack of previously drilled outer holes provides a clean and direct path toward the block walls. Once the block is prepared accordingly, the splayed caps must then be fit and machined like any other aftermarket cap.

Four-bolt main caps with angled outer bolts draw the block toward the crankshaft centerline at different angles. The bolts extend into a thicker portion of the main saddle where it ties into the outer walls improving overall block rigidity. Splayed main caps are a popular upgrade when using a stock Pontiac block in a high-performance application.

Pontiac used two-bolt main caps in most of its production engines. Although some blocks were drilled and tapped for four-bolt caps, most were originally fitted with two-bolt units like this. They are sufficient up to about 600 hp. Honing improves alignment and bearing life without grossly altering the position of the crankshaft centerline.

Four-bolt main caps were used at the center three journals in high-performance applications. The four-bolt cap is larger than a two-bolt unit, and serves to improve main saddle rigidity and overall durability, particularly with 3.25-inch main journals. Stock Pontiac caps were constructed of a cast-iron alloy. Installing used caps onto another block requires proper machining.

Pro-Gram Engineering produces top-quality aftermarket main caps for the Pontiac V-8 constructed of billet steel. They're ideal for high-performance applications where additional cap strength is required. Though Pro-Gram produces all five main caps for 3- and 3.25-inch blocks, the center three are retained by four straight bolts or with angled outer bolts. The block and caps must be machined accordingly.

one inch below the large expansion plug holes in the side.

A variety of water jacket fillers are on the market today. I recommend premium filler such as Hard Blok or that from Ken's Speed & Machine Shop. Following the supplied mixing and installation instructions provides the best results.

Main Caps

The center three main journals of a Pontiac V-8 endure the greatest load, and the corresponding bearing caps were generally retained by only two bolts early on. Pontiac determined that using beefier main bearing caps and four retaining bolts in the number-2, -3, and -4 journals improved rigidity and reliability, especially in applications with 3.25-inch-diameter main journals, where more of the main cap is lost to journal area. Factory four-bolt caps are commonly found on certain Ram Air, H.O., and Super Duty engines of the 1960s and 1970s. Most Pontiac blocks during that era were machined to accept two-bolt caps only, but 455 blocks produced through about 1974 were mostly machined for four-bolt caps even though two-bolt units were installed most often.

Stock Pontiac main caps are quite durable and adequate for power levels approaching 550 hp. Factory four-bolt caps can likely sustain even greater power levels. At about 550 hp aftermarket four-bolt main caps should strongly be considered for the center three journals. Installing four-bolt caps isn't as easy as bolting them on, however. A block and its main caps are a precision assembly that's machined as a unit. Whenever a different main cap is used, whether a used original from another block or a new aftermarket unit, it must be precisely machined. That can include align-boring, thrust machining, and/or possibly relocating or resizing the dowel holes.

Billet-steel main caps are very strong and a few different aftermarket companies offer Pontiac V-8 units. Lesser-quality steel caps can be too hard, which makes them extremely difficult to machine. In those instances, the line-bore cutting equipment can bounce erratically, removing valuable material from the

A main cap must offer some absorptive quality to prevent main saddle fatigue. Milodon's ductile-steel main caps are an excellent high-strength choice for such applications. Only available for the center three journals and with four straight bolts, they install and machine easily and offer the durability required for the most extreme applications.

block in the process. Once installed, the main caps become an integral part of the block and improve rigidity while absorbing crankshaft shock load. Lesser-quality billet caps can also be so hard that they resist shock load, leaving the block to absorb it all. Over time the block can fatigue and fail.

I feel the best billet main caps available today for Pontiac V-8s are produced

What is a Dry Deck Engine?

A very high compression ratio, spraying nitrous oxide, or the boost associated with forced induction can cause cylinder pressure and temperature to spike under load. As pressure peaks, it essentially attempts to separate the heads and block. That not only greatly stresses the cylinder head fasteners, it can also compromise the head gasket's ability to maintain sufficient seal. Once the gasket loses that ability, coolant can slip past the gasket and into one or more cylinders, causing the engine to hydraulically lock. Complete and catastrophic failure is often the result.

In such applications it's sometimes best to eliminate the coolant path between the block and cylinder heads. The practice, best known as "dry decking," consists of physically blocking those passages to improve reliability. Generally speaking, the coolant holes in the opposing block and cylinder head deck surfaces are tapped to accept threaded plugs and/or welded closed before machining. A secondary path for coolant circulation usually includes some type of hose or tubing that reconnects the block and heads.

While there are many benefits to a dry-decked engine for specialized applications, and is a modification I recommend for an engine that's continually subjected to extreme cylinder pressure, I don't necessarily recommend dry-decking for a dedicated street engine. It may inhibit proper engine cooling and can be costly and impractical to maintain. It's certainly possible, however, and a well thought-out flow path, a high-volume water pump, and a very efficient radiator are ways of ensuring its success on the street.

If you feel that dry-decking may improve the reliability of your Pontiac, plan to discuss it with your Pontiac engine building specialist. Oftentimes, specific head gaskets and special block machining can further enhance the effect, allowing the engine to tolerate even more cylinder pressure for even greater performance. If your build includes an aftermarket Pontiac block, be sure to discuss dry-decking with your vendor before making the purchase. Certain blocks may be delivered with an undrilled deck surface, which can save you time and money. Coolant holes can always be added later.

A "dry decked" block lacks the passages that allow coolant to circulate between the block and cylinder heads. It's desirable for applications where extreme cylinder pressure exists and a coolant system breach is possible. Dry decking a stock Pontiac block requires the installation of pipe plugs and/or welding, but certain aftermarket blocks are delivered with a dry deck. If a wet deck is desired, coolant passages can be drilled accordingly using a typical head gasket as a template.

by Pro-Gram Engineering. They're strong enough to add significant benefit, yet machine relatively easily and exhibit excellent dampening ability. Pro-Gram offers complete main caps for all five journals in 3- and 3.25-inch diameters. While the front and rear caps are sold individually, the center three are sold as a set and are designed for four-bolt retention only, with straight or splayed outer bolt options. Pro-Gram main caps sell for about $300 and can be purchased directly from the manufacturer or from your favorite Pontiac vendor.

Milodon has been producing high-quality ductile steel main caps for the center three journals of Pontiac V-8s for several years. Retained by four straight bolts, the caps are very strong and install relatively easily. An added benefit is that the thrust surface on the number-4

CHAPTER 2

Whether cast-iron or aluminum, aftermarket Pontiac blocks are designed to allow for easy installation into any chassis. Each side contains the appropriate bosses to accept most original engine mounts. It also accepts the original oil dipstick tube that's driven into the block. Coolant jacket plugs are press fit or screw in, depending upon the manufacturer.

There are a few different aftermarket Pontiac V-8 blocks presently available that are ideal for applications approaching 750 hp and beyond. AllPontiac.com manufactures two distinct units. The IA-II is a cast-iron block that's been on the market for several years, while the all-new aluminum block was a recent release. Both generally boast beefier main saddles and four-bolt main caps to significantly improve bottom end rigidity, particularly when using a very long stroke aftermarket crankshaft measuring 4.5 inches or more to produce a maximum displacement of 541 ci.

cap is the same diameter as the block, which means less machine work when compared to others. Expect to spend around $300 for a set.

Aftermarket Blocks

Though complete failure with a stock block is rather rare when output is less than 750 hp or so, there are distinct applications where an owner is looking to grossly increase displacement, or the intended performance level surpasses the capability of a stock design. The cost to prepare a stock block for a race application is nearly as much as the aftermarket Pontiac V-8 blocks available today. New blocks that contain additional material in critical areas are available for hobbyists looking to go to the next performance level, and are an excellent investment if future performance modifications are planned.

Two separate companies presently produce aftermarket Pontiac V-8 blocks. Each contains external dimensions similar to an original Pontiac unit and will accept many original Pontiac pieces, but are generally much beefier throughout. AllPontiac.com produces its cast-iron IA

K&M Performance produces two aftermarket Pontiac V-8 blocks. They are nearly identical, except the MR-1 is constructed of cast iron, while the MR-1A is constructed of cast aluminum. A maximum displacement of 541 ci is possible. The lifter galley is heavily reinforced to accommodate the most radical camshaft and the main saddles are much thicker throughout.

II block and an all-new aluminum casting. K&M Performance produces its MR-1 in cast iron or cast aluminum. Available from many Pontiac vendors, the aftermarket blocks are an excellent foundation for most max-performance applications, especially where displacement of 541 inches, and possibly more is desired.

The IA II block is constructed of high-nickel iron and features siamesed

The aftermarket blocks from AllPontiac.com and K&M Performance feature steel main caps and high-quality main studs. The center three caps are designed for four-bolt (or studs in this case) retention with outer-holes angle for maximum block rigidity. The main caps are drilled to accept a stock windage tray and/or oil dipstick tube.

Camshaft bearings generally are not subjected to the same pressure as other engine bearings. The stock replacement Pontiac units available from many sources are usually more than sufficient for the moderate spring pressures associated with hydraulic and solid roller camshafts. Cam bearings must be installed using the proper equipment or damage can result.

Dura-Bond offers a high-quality Pontiac camshaft bearing set that features a dry-film lubricant coating. The coating reduces friction in applications where extreme valvespring pressure can damage standard bearings, such as with a very aggressive solid roller camshaft with high-valve lift.

cylinders that tolerate a bore diameter up to 4.4 inches while maintaining a wall thickness that's nearly .300 inch thick. The deck surface measures .750-inch thick and is delivered "dry." The coolant passages must be drilled for "wet deck" applications. A tall-deck option that increases compression distance for specialized applications is also available. The lifter galley is greatly reinforced and is designed to accommodate the very long duration and high-lift camshafts that can damage or destroy the lifter bores in a stock Pontiac block.

Main journals are available in 3- or 3.25-inch diameters. Billet-steel main caps are standard, and the center three are a four-bolt design with angled outer bolts. All caps are registered to the main saddles and use larger 3/8-inch dowel pins for positive location to prevent walking, which can negatively affect oiling and bearing alignment at high RPM. The camshaft oiling passages that extend upward from the main journals are drilled smaller to strengthen the center of the block and prevent cracking in extreme conditions

Main registering creates an oil pan rail that's .125 inch deeper than stock. It improves block rigidity, but also requires a slightly longer oil pump driveshaft, which AllPontiac.com supplies. The IA II block comes with threaded freeze plugs and a bellhousing flange containing two distinct bolt patterns for both Buick-Olds-Pontiac or Chevy-type transmissions. It also features five mounting holes per side to accommodate any original chassis motor mounts. At about 250 pounds with main caps, the IA II block weighs 40 to 50 pounds more than a race-prepped stock block, but the extra weight is well worth it—the IA II is extremely durable. Expect to spend around $3,000 for a basic IA II block. A wide variety of options are available at extra cost.

AllPontiac.com has recently developed and released a completely new lightweight aluminum Pontiac block for those most serious about performance. While containing many of the same design features and options as the cast-iron IA II, the high-quality aluminum casting is even beefier than its iron counterpart. It weighs nearly 140 pounds with the main caps, or about 115 pounds less than the cast-iron piece, but with a near-equal amount of durability. Expect to pay nearly $5,000 for the standard block. Many extra-cost options are available.

K&M Performance produces the cast-iron MR-1. Many mistakenly believe that it's a copy of the IA II, but the MR-1 was actually designed to eliminate the weaknesses associated with the stock Pontiac block in max-performance applications while accepting as many stock Pontiac pieces as possible. The MR-1 allows for a bore diameter up to 4.4 inches while maintaining a wall thickness of .200 inch. The deck surface is very thick and is drilled for "wet deck" applications. The coolant passages are sized to accept common pipe plugs after tapping. The lifter galley is heavily reinforced.

The MR-1 is available with 3- or 3.25-inch main journals. Billet-steel main caps are located using typical 5/16-inch dowel pins. The center three are retained by four bolts and the outer most bolts are angled for better block integrity. Like a stock Pontiac block, the MR-1 uses pressed freeze plugs and the oil pan rail is in its original position, which allows the use of many typical off-the-shelf Pontiac parts.

A typical MR-1 block weighs around 250 pounds and features five mounting holes per side for ease of installation. The base block sells for around $3,000 and many extra-cost options are available. K&M Performance states it's capable of safely enduring up to 2,500 hp or slightly more.

Shortly after introducing the cast-iron MR-1, K&M Performance developed and released an aluminum variant known as the MR-1A. Essentially a cast-aluminum copy of the iron MR-1, the blocks share the same design improvements and tooling. At a svelte 125 pounds, the MR-1A weighs about half as much as its cast-iron brethren, but because aluminum isn't nearly as durable as iron, K&M Performance presently recommends it for applications up to 1,500 hp. Pricing starts around $4,500 and many extra-cost options are available. For the most serious racer, K&M Performance also produces a billet-aluminum Pontiac V-8 block on a custom-order basis. Expect to spend around $10,000 for what may be the strongest Pontiac block ever available.

Camshaft Bearings

Camshaft bearings are not required to carry very heavy loads. Generally a one-piece, two-layer design, it includes a steel shell and a very soft overlay. ACL, Clevite, Dura-Bond, Federal-Mogul, and King produce quality stock-replacement cam bearings suitable for most applications.

Dura-Bond offers a performance bearing that better resists the high valvespring pressures associated with aggressive roller camshafts. It even offers an optional dry-film lubricant coating intended to reduce friction in extreme applications. While most cam bearing sets come in standard size, Dura-Bond also offers a .010-inch oversize for instances where honing the camshaft tunnel is necessary.

Specific camshaft bearing sets with needle rollers are available for other makes. They can be adapted for use in the most severe Pontiac applications where the highest camshaft loads exist. The block's camshaft tunnel must be bored accordingly to accept the oversize bearings, however, and that can remove valuable material within the main saddle of a stock Pontiac block, possibly compromising integrity. Roller cam bearings are a more feasible option when using a beefier aftermarket block for a build.

Rear Main Seal

Pontiac originally used a woven rope constructed of graphite-coated asbestos to seal its V-8's rear main journal. The

The rear main seal groove machined into the block and cap features drilled holes to anchor the original rope main seal and prevent it from rotating during normal operation. These holes must be filled with silicon sealer to prevent oil leaks when using an aftermarket lip-type seal. Aftermarket blocks and main caps do not contain these anti-rotation holes, so a small roll pin is required when using an aftermarket rope seal. Also note that the seal diameter varies among 3- and 3.25-inch blocks.

two-piece seal was packed into a corresponding groove machined into the block and main cap. The seal grooves had cavities drilled in them. As the rope was packed firmly into place during installation, it filled those cavities, effectively anchoring it and preventing rotation during operation. Excess portions of rope were then trimmed away, and the main cap installed.

Asbestos was chosen for strength, with a graphite coating to reduce friction. The original rope seal worked quite well

BLOCKS

Pontiac originally used a rope seal constructed of woven asbestos (bottom), and it performed quite well. Modern replacements are constructed of woven fiberglass or Kevlar (top) and can be much more difficult to work with. Unless you have a supply of NOS originals, avoid the modern replacement because severe leaks can result even with careful installation.

A lip-type seal like this from BOP Engineering is an excellent design that consists of a Viton-coated steel core. The lip is designed to remain in constant contact with the crankshaft's seal surface, essentially eliminating leaks. It's compatible with the serrated lubrication grooves found on stock-type cranks, but journal polishing can be necessary.

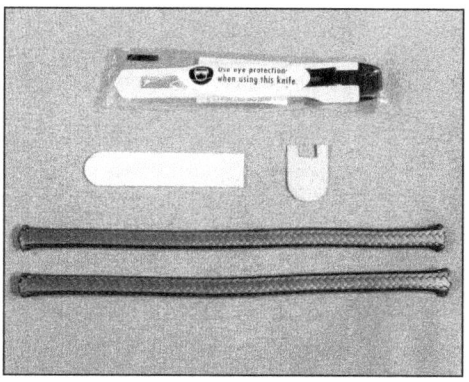

The GraphTite seal from Best Gasket is a modern rope replacement that's constructed of graphite-coated Teflon strands. It fits and functions like an asbestos original and can be relatively leak free when installed as directed. It's an excellent choice, but I recommend using it only with crankshafts that contain serrated lubrication grooves.

for many years. It was lubricated during normal operation by small grooves on the crankshaft contact surface. While some oil "wicking," or seepage that causes an occasional drip can be considered normal, larger leaks that result in small puddles were rather uncommon. New OE-spec replacement seals, such as those from Fel-Pro, were also constructed of asbestos and rarely problematic if installed as Pontiac originally instructed.

Government legislation eventually restricted the use of asbestos and companies were forced to find new materials for rope seal construction. Fel-Pro's solution was braided fiberglass with a graphite coating. When compared to an original asbestos unit, the fiberglass replacement was less flexible and slightly more difficult to install, but it was effective.

Fel-Pro changed the base material of its rope seal from fiberglass to Kevlar in 1998. Since Kevlar better tolerates heat, Fel-Pro felt it would improve longevity. Though the recommended installation is identical to that of its predecessors, the Kevlar seal is even more difficult to work with than fiberglass. It's much firmer and that makes it even more difficult to pack and trim accordingly. In fact, it's not uncommon to use several fresh razor blades trimming a single seal during an installation.

Kevlar rope seals are generally included in every modern Fel-Pro engine gasket set. While I feel most other Fel-Pro gaskets offer the best functionality, the modern rope seal is one to avoid. It isn't very forgiving and can leak severely even with careful installation. Though used successfully on occasion, most report poor results. Fel-Pro reports it has no plans to deviate from the current design since it's as close to OE-spec as modern regulations allow. That has forced many hobbyists and professional engine builders to resort to other alternatives.

A lip-type rear main seal has been commonly used by various other auto manufacturers for many years. It's generally constructed of a temperature-resistant rubber material such as neoprene and the "lip" is a formed "flap" that remains in constant contact with the sealing surface on the crankshaft.

The rubber seal originally designed for certain large-cube Cadillac V-8s is a shape and size similar to that required for a Pontiac. Many hobbyists have successfully retrofited a Cadillac seal into their Pontiac, but installation is rather complex. Careful installation does yield positive results in most cases but anything less can result in a significant leak.

In 2000, BOP Engineering introduced a new molded lip-type seal specifically for the Pontiac V-8 and it proved to be an immediate success. It is constructed of a hardened steel core coated in Viton, a pliable material that's oil and temperature resistant. It features a multiple-lip design that's intended to improve oil control.

A lip-type seal requires near absolute concentricity for maximum sealing ability. Pontiac didn't hold the depth of the rear seal groove to any particular tolerance during machining, however. Since the rope seal was pliable, there wasn't a great need for the groove to be precisely concentric to the crankshaft centerline.

BOP suggests a variance of no more than .010 inch from the crankshaft

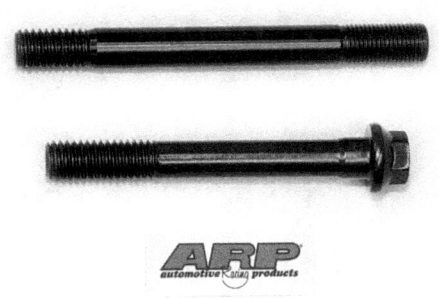

ARP offers specific bolt kits for many Pontiac cylinder heads including cast-iron originals and popular aftermarket units. It also offers a cylinder head stud kit for max-performance applications. If ARP doesn't offer a specific fastener kit for your particular cylinder heads, it can assemble a custom set to accommodate your needs.

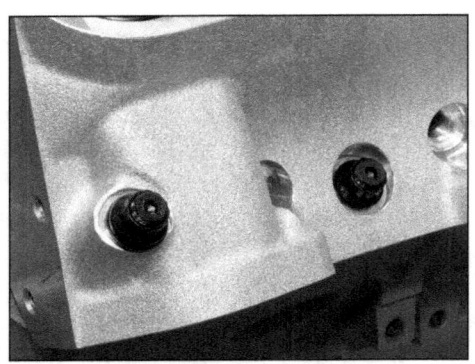

Cylinder head studs reduce the amount of twisting force the block sees during tightening and can provide a clamping load that's more consistent. I recommend ARP cylinder head studs when performance exceeds 650 hp. ARP's specific moly-based thread lubricant on the threads and under the washers during install yields the best results.

Four-bolt main caps improve main saddle rigidity, but angled outer bolts draw the entire block together. "Splayed" main caps significantly make the outer block walls an integral part of the main saddle, improving overall block strength. An original two-bolt Pontiac block can easily be modified to accept splayed caps. Aftermarket Pontiac blocks feature splayed caps with main studs as standard equipment.

Pontiac main cap bolts were quite reliable, but modern fasteners may be a better option as the originals are 30 to 40 years old. ARP offers complete stud kits specifically designed for two- and four-bolt Pontiac main caps. In addition to starting with a new fastener, the studs eliminate the twisting force that main saddles otherwise see when you tighten the bolts. A stock windage tray and the oil pan can sometimes interfere with the main studs during assembly, but it can be easily corrected.

centerline for proper operation of its seal. I've found that closely following BOP's installation instructions provides excellent results. A similar seal from Tin Indian Performance is also available, and though I have no direct experience with it, I suspect it performs as well.

A new Pontiac V-8 rope seal was introduced around 2007. The GraphTite seal from Best Gasket is constructed of braided Teflon that's coated with graphite. It fits and installs like an asbestos original, and comes with detailed instructions, cutting blade, and template. More recent kits include a small roll pin to positively locate the seal when used with the aftermarket Pontiac V-8 blocks and caps, which lack the anti-rotation holes of an original. I have found it provides an excellent seal.

It's best to use one of the aftermarket rear main seal options mentioned above. The type that's best for you is often personal preference, but it can also depend upon the concentricity of the seal groove in the block and main cap. I suggest physically measuring the concentricity of your Pontiac block before deciding which rear main seal to use in your project.

Fasteners

When developing its V-8, Pontiac used 1/2-inch-diameter cylinder head bolts for maximum retention. An original Pontiac bolt in excellent condition can be reused several times, but there's sometimes no way of knowing how many times it's been previously used. Reproduction head bolts are available and the quality can range from very good to very poor. Investing in high-quality aftermarket cylinder head fasteners is the best choice.

Automotive Racing Products (ARP) produces a variety of premium automotive fasteners. It offers complete sets containing bolts in the proper lengths required for original Pontiac D-port and round-port cylinder heads as well as the aftermarket Edelbrock offerings. Custom sets for other aftermarket cylinder heads are also available. It's best to check with the cylinder head manufacturer to determine exactly what's required.

When installed as ARP suggests, using its specific moly-based thread lubricant, you can expect excellent results. I strongly suggest measuring the overall length of each bolt and its corresponding hole in the block to be sure the bolt doesn't bottom out and provide a false torque reading during installation. This is especially important if the deck surface of the block and/or cylinder heads has been machined by any amount.

When approaching the 650-hp level, it may be best to consider ARP's cylinder head stud kit as opposed to bolts for improving retention. The stud is threaded into the block and a nut and washer assembly applies the clamping load. The stud eliminates the twisting that otherwise occurs when threading a bolt into the block, and it stretches more smoothly when torque is applied to the retaining nut. It reduces block distortion and improves clamping consistency and thread life.

The fasteners that Pontiac used to secure the main caps are of the highest quality. Those bolts are, however, more than 40 years old in most instances. If you or your machinist feels new fasteners are necessary, ARP offers a complete stud kit for two-bolt and four-bolt Pontiac V-8s.

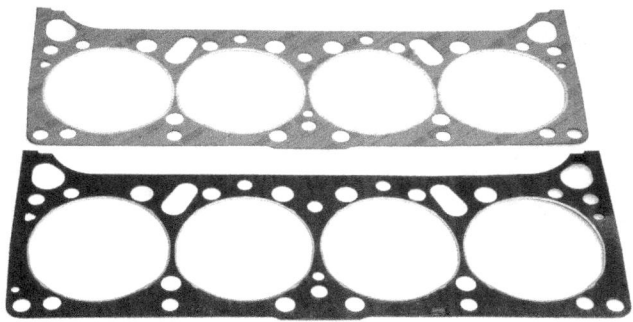

Fel-Pro offers two different Pontiac cylinder head gaskets that share the same basic shape and size. The design is a compilation of the many stock replacement units Fel-Pro offered over the years, and it's compatible with most popular Pontiac blocks. PN 8518 (top) is a stock replacement unit while PN 1016 (bottom) is a high-performance gasket with a solid-steel core and premium fire rings. Completely coated in Teflon, Fel-Pro recommends installing its head gaskets dry.

A drawback to Fel-Pro's cylinder head gasket is that its coolant holes are rather large and located fairly closely to the combustion ring. It's quite visible when overlaid on an NOS original Pontiac head gasket (shown here). If the ring is compromised in any way, compression can push past the Fel-Pro gasket and enter the coolant system, which can result in major damage. While it may not be a concern for applications producing 700 hp or so, it can be an issue with forced induction or an extreme compression ratio.

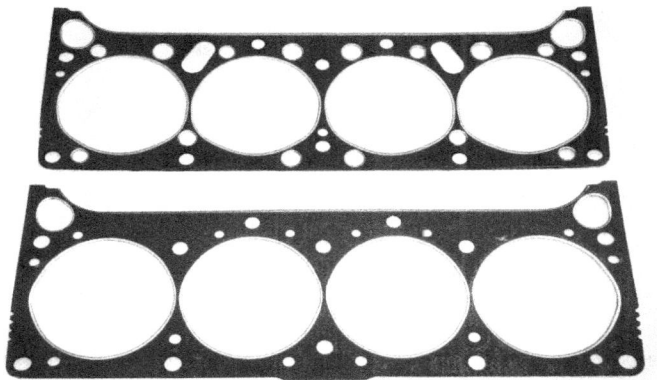

Composition cylinder head gaskets from Butler Performance were developed using the original Pontiac blueprints. This design consists of a steel shim with a graphite coating on one side, and can provide excellent seal when combined with copper sealer. Butler Performance was able to size and locate bore-diameter and coolant passages for specific applications just like the originals. It's easy to see how Butler's 421-spec unit (top) differs from the 400/455 unit (bottom).

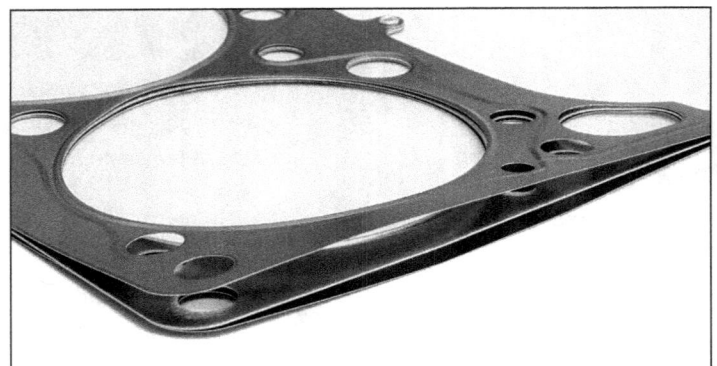

MLS gaskets are the ultimate in cylinder sealing. The multiple layers allow for varying degrees of thermal expansion while providing maximum sealing capability in applications where extreme cylinder pressure exists. MLS head gaskets are commonly used by all auto manufacturers in modern production vehicles and are an excellent choice

Cometic manufactures premium MLS cylinder head gaskets. It offers a variety of options for traditional Pontiac V-8 blocks as well as the aftermarket offerings with various bore diameters and compressed thicknesses. Its gaskets are an excellent choice for maximum-performance applications, and are completely compatible with stock rebuilds too.

The main studs offer the same advantages as cylinder head studs, and that includes less twisting on the block, clamping load consistency, and improved thread life. ARP main studs can be used in conjunction with stock main caps at any point, but are required with aftermarket main caps, which may or may not include new fasteners for installation.

Head Gaskets

A cylinder head gasket is designed to keep combustion heat and pressure within the cylinder and prevent coolant that passes between the block and head from leaking outward or into the cylinders. Should combustion pressure blow past or burn through the gasket's combustion armor, escaping compression generally leads to complete gasket failure, and that can lead to significant engine damage if not detected in time.

The head gasket Pontiac used for its V-8 was a composition construction consisting of a coated steel core. Many companies produced stock-replacement Pontiac head gaskets over the years, but few were as good those originally supplied by Pontiac.

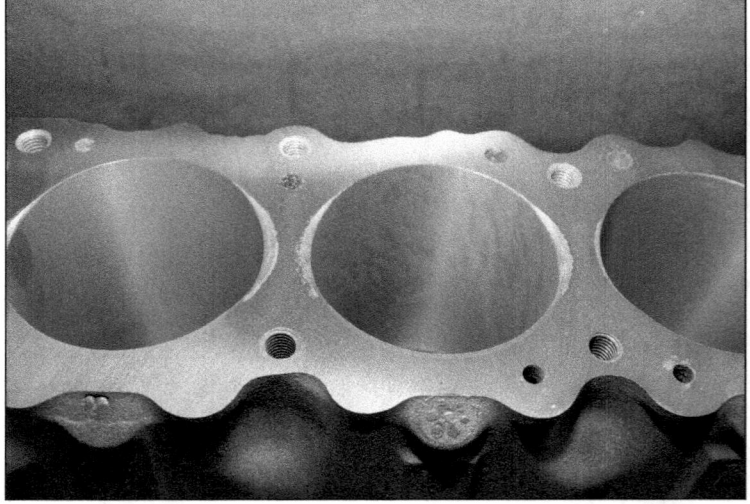

When cylinder heads and blocks are surfaced, the equipment leaves behind a finishing pattern. The roughness of the finish must be compatible with the cylinder head gasket or it can be torn apart as the block and heads expand and contract during normal operation. Most gasket manufacturers publish a roughness average (RA) range for their gaskets. It's very important to follow the RA recommendations for maximum gasket reliability.

Fel-Pro introduced PN-8518 for the Pontiac V-8 during the 1970s. Containing a bore diameter of 4.3 inches and a pre-flattened thickness of .041 inch, it's a stock-replacement design that's compatible with most Pontiac blocks. Constructed of a perforated steel core with a Teflon outer coating, it features tin-plated steel combustion armor. PN-8518 head gaskets sell for about $60 per pair and are an affordable choice for any stock rebuild, and even performance applications with a compression ratio up to 10.5:1, or slightly more.

A second Pontiac offering appeared in the Fel-Pro catalog during the early 1980s. PN-1016 is specially designed for high-performance applications with high compression and/or combustion pressure. Constructed of a Teflon-coated solid-steel core, it features stainless-steel combustion armor and a wire ring combustion seal for maximum combustion containment. It contains a chamfered bore diameter or 4.3 inches and pre-compressed thickness of .039 inch. Selling for less than $80 per pair, it's a good choice for naturally aspirated engines with a maximum compression ratio of 12:1 and/or mildly boosted applications.

In the early 1990s Butler Performance (BP) began offering its own composition cylinder head gasket using original Pontiac blueprints. It features a steel core with graphite coating on one side. BP revised the coolant passage sizes to closer match those in the block to improve reliability. Available with various bore diameters, it compresses to about .045 inch after installation. BP recommends a light coating of copper sealer on the steel side for maximum block seal. BP's composite gaskets are very durable and a popular choice for compression ratios up to 11.5:1 or moderately boosted engines. A pair sells for less than $50.

Copper head gaskets have long been the only choice for max-performance engines with extreme cylinder pressure and/or combustion heat. The durable copper construction prevents the blowouts and/or burning that can occur when exposing composition gaskets to high levels of nitrous oxide or boost. It's often combined with a wire O-ring, which resides in a machined groove in the block and prevents combustion pressure from blowing past the gasket. A copper gasket's main drawback is its inefficiency to prevent coolant or oil leaks. Liberal coats of surface sealer during installation were once required, but SCE Gaskets offers top-quality copper gaskets with built-in seals, and it's an excellent concept. Copper gaskets are available from several companies and remain the best choice for extreme performance applications.

Multi-layer steel (MLS) gaskets are today's choice for most max-performance applications. As the name implies, an MLS gasket is constructed of multiple layers of steel gaskets with a tough-yet-pliable exterior coating. The multi-layer design improves sealing by spreading the load across the entire gasket surface while the outer coat seals the gasket against most types of mating surface or load. MLS cylinder head gaskets have proven to be as durable as any copper unit in all but the most extreme applications, but they lack the sealing issues that plagued racers for years.

Cometic is the only company that presently produces MLS cylinder head gaskets for the Pontiac V-8. Bore diameters range from 3.75 to more than 4.4 inches and standard compressed thickness measures .040 inch. Any other bore diameter and thickness from about .020 to .120 inch is available on a custom-order basis. Suitable for stock rebuilds and max-performance applications alike, the only drawback to an MLS gasket is cost. Expect to spend around $200 per pair, but it's an excellent value considering its durability.

When selecting cylinder head gaskets for your Pontiac, cost and application are among the most important considerations. The best choice is the most affordable gasket that operates reliably in the intended application. Compressed thickness is also important because it has a direct effect on static compression ratio. Each .010-inch difference of head gasket thickness alters the static compression ratio by about .15:1 on a typical 467-ci Pontiac. Certain cylinder head gaskets are also reusable. Assuming they are not stuck to the block and do not tear during disassembly, Fel-Pro and Cometic gaskets can be reused with the same block and cylinder heads at least a few times.

The most import aspect of head gasket choice is the materials it is used with. Aluminum expands about twice as much as cast iron when exposed to the same heat level. A bimetal engine, or one that uses an iron block and aluminum cylinder heads, can literally scrub a head gasket apart as it reaches normal operating temperature if the opposing surfaces have a machining finish that's too rough. That scrubbing can eventually lead to gasket failure.

Pontiac's original head gasket was very forgiving and conformed to most finishes. The stock-type gasket from Butler Performance performs similarly.

Bolt-On Components

There are two components commonly found within the oil pan of a high-performance Pontiac V-8. A windage tray is a baffle that bolts to the main caps or is an integral part of the oil pan. The aeration created within the crankcase as the crankshaft rotates at high RPM is known as "windage." As oil drains toward the bottom of the engine, it can splash when it reaches the sump. Windage can keep smaller droplets of splashing oil in suspension, which can then collect on the crankshaft and connecting rods, increasing parasitic drag and adversely affecting performance.

Though the mid 1970s, Pontiac used a stamped-steel baffle bolted to the main caps to reduce windage in its V-8. It doubled as a means of retaining the lower dipstick tube within the block. Reusing an original Pontiac baffle can be dicey, however. After years of enduring engine harmonics, the aged sheet metal can fatigue and crack during use unbeknownst to the owner. It can eventually lead to complete baffle failure, which can then cause serious and possibly even catastrophic collateral engine damage.

The aftermarket Tomahawk windage tray is an excellent replacement. A copy of the stock Pontiac unit, it's constructed of much thicker metal and is improved slightly. With its relatively reasonable cost and ease of installation, I strongly suggest a Tomahawk unit as opposed to reusing any original. Canton and Milodon also produce high-quality windage trays for the Pontiac V-8. The two are uniquely different and, unlike the stock piece, and may better serve a max-performance application. AllPontiac.com also offers its own windage tray for its aftermarket blocks.

A crank scraper is most commonly found within an oil pan of a race enigne. It's a length of flat metal that's installed very closely to the crankshaft and connecting rods on the passenger side of the block, and "scrapes" oil from them during rotation to keep excess oil off the cylinder walls and reduce parasitic drag at high RPM. Installed clearance is generally between .020 and .060 inch depending upon the application. Due to variances in crankshaft counterweight and connecting rod design, a crank scraper will require some type of custom fitting and requires extra preparation to seal properly. Crank scrapers are available from many vendors and sell for less than $50.

Some feel that with a crank scraper, a windage tray serves no purpose. Others elect to run both simply to improve oil control and ensure their Pontiac V-8 is as free of parasitic drag as possible. I believe at least one should be used in any performance engine and see no harm in using both if applicable. Whether together or singly, the reduction of parasitic drag associated with such components can improve high-RPM performance by as much as 5 to 7 hp. Discuss with your engine builder whether any or either is required at your performance level.

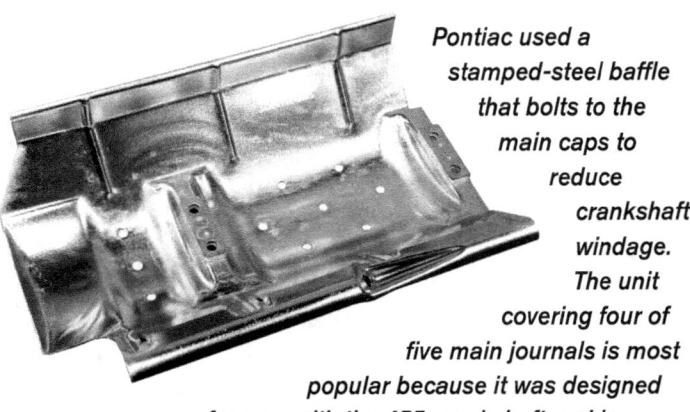

Pontiac used a stamped-steel baffle that bolts to the main caps to reduce crankshaft windage. The unit covering four of five main journals is most popular because it was designed for use with the 455 crankshaft and is generally a bit thicker. Years of exposure to engine harmonics can cause fatiguing and cracking. I much prefer the Tomahawk windage tray from Pacific Performance Racing. It's a stock replacement 4/5 unit that's lightly modified to improve performance and reliability and installs just like the original.

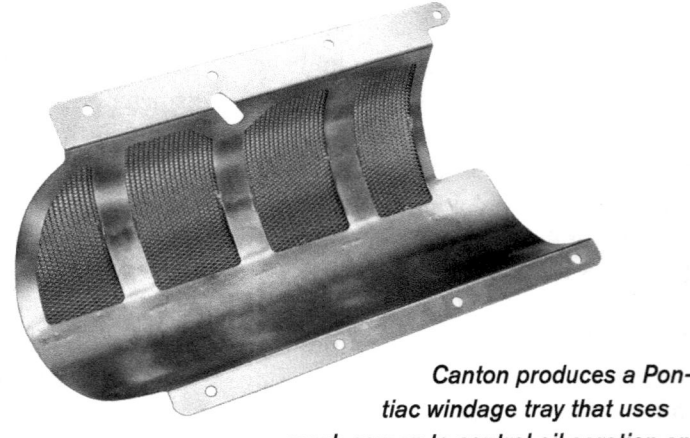

Canton produces a Pontiac windage tray that uses mesh screen to control oil aeration and improve horsepower. Covering all five journals, it rests on the oil pan rail and is retained by the oil pan and its fasteners. It's compatible with the stock oil pan and most stroker assemblies. Canton also offers a similar unit that uses stamped louvers in place of the mesh screen.

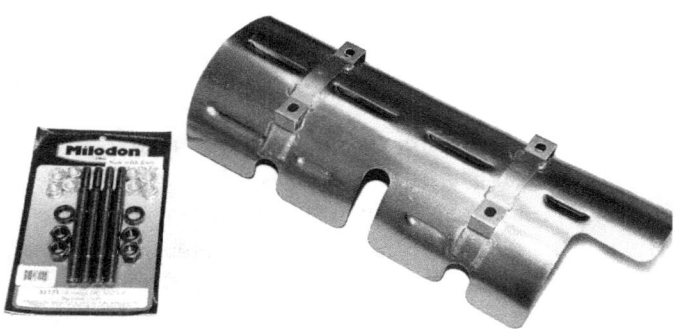

Milodon offers a unique stamped-steel windage tray for the Pontiac V-8. It requires the use of specific main cap studs which Milodon supplies. While the tray is compatible with the stock Pontiac oil pan and most stroker assemblies, best results occur when combining it with a Milodon oil pan.

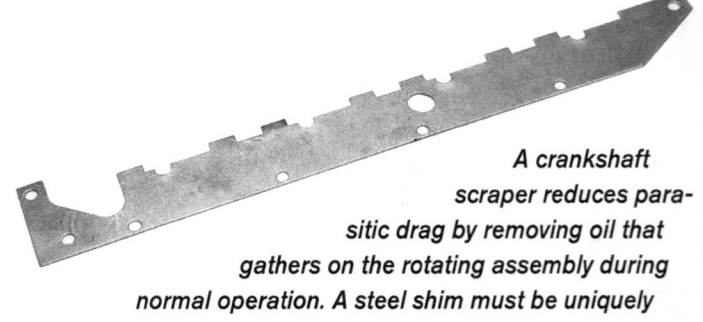

A crankshaft scraper reduces parasitic drag by removing oil that gathers on the rotating assembly during normal operation. A steel shim must be uniquely trimmed for each particular application. This unit is available from Ken's Speed & Machine Shop and is precut for the Pontiac V-8. It requires only minimal trimming to gain sufficient crankshaft and connecting rod clearance.

CHAPTER 3

CRANKSHAFTS

A crankshaft transfers the energy created during combustion into a rotating force that's channeled through a transmission and rear axle, and finally to the tires for vehicle motivation. As each cylinder fires, the crankshaft must be rigid enough to endure the shock loads associated with combustion, but flexible enough to not fatigue and fail at any reasonable engine speed. It's a delicate balance that requires precise balancing and exact tuning to prevent premature failure.

Original Pontiac Cranks

When Pontiac developed its V-8, the crankshaft was engineered to provide a long and reliable service life. It's supported by five large main journals and uses large counterweights to maintain centrifugal balance and motion during normal operation. The design features large bearing surfaces and a considerable amount of overlap between the main and rod journals to increase rigidity. Forward/rearward thrust is taken up at the fourth main journal.

Enlarging bore diameter and adding crankshaft stroke length were ways that Pontiac increased the displacement of its original 287 to an eventual size of 455.

Pontiac used a forged crankshaft in its early V-8s. Castings first appeared in 1959 and were used through the end of V-8 production in 1981. Cast Pontiac cranks are quite durable and can offer excellent reliability to 600 hp or more with typical preparation. Examples from 1965 through the mid 1970s offer the best "swap-ability" and are generally the most compatible with popular aftermarket components.

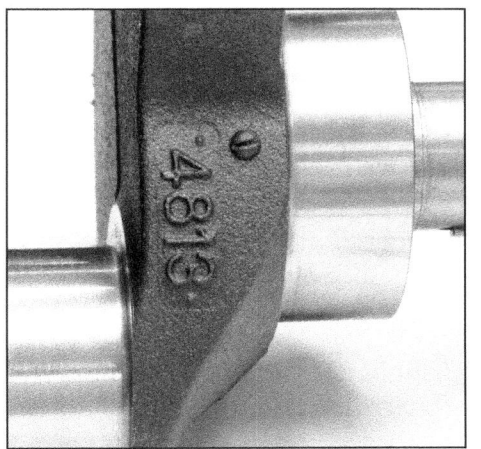

Casting numbers can be used to identify most Pontiac crankshafts. Location varies, and I have found original Pontiac cranks without part numbers, some partially or fully stamped, and others with full numbers. This particular casting was used for mid-1970s 350 and 400 applications, and contains only the first four digits of the part number. Since the two engines use the same crank but require different balances, the remaining two digits were stamped elsewhere. It makes identification very difficult.

HOW TO BUILD MAX-PERFORMANCE PONTIAC V-8s

Main journal size initially measured 2.5 inches, but to maintain sufficient journal overlap for maximum rigidity as stroke lengthened, it increased to as much as 3.25 in later years. The crank pin, or connecting rod journal, as it's often called, remained at 2.25 inches throughout the entire production run.

Pontiac cranks are generally categorized and referred to by main journal sizing. The 3-inch unit was used in all V-8s up to 400 inches produced between 1959 and 1979. The 3.25-inch unit was introduced in the early 1960s for the 421, and is found in all other Pontiac engines displacing up to 455 ci. Generally speaking, cranks with the same main journal sizing freely interchange, but there are some slight differences that can affect it (covered later in this chapter).

All Pontiac cranks produced though 1958 were constructed of forged steel. Though quite durable, they're of no real use to performance enthusiasts today because the main journals measure less than 3 inches. Bearing spacers that allow the use of such cranks in 389 and 400 blocks were once available from a few sources, and can still be made by most competent machine shops. With so many drop-in forged-steel cranks available today, however, there's less of a need to use an original forging.

Once Pontiac increased main journal diameter to 3 inches, it found that a cast crank could offer sufficient durability and reliability for its production V-8, allowing the elimination of the costly forging in all but a limited number of highly specialized competition engines. Pontiac began casting its production cranks in 1959. It used pearlitic malleable iron (PMI), Armasteel, or nodular iron, depending upon the model year and application. The castings proved to be very rigid and resisted flexing.

Despite the material it's constructed of, the stock castings contain about the same tensile strength. Those constructed of Armasteel, a GM trade name for a specific iron alloy that exhibits some steel-like qualities has the highest, but only by a slight margin. The nodular iron unit was used in most every Pontiac engine during the years when production peaked. Such cranks are definitely the most common and are nearly as durable as any Armasteel unit. A nodular iron casting is identifiable by a large "N" on or near the first counterweight.

Common Pontiac Crankshaft Specifications

Pontiac used many different part numbers for its crankshafts. This is a listing of those I am most familiar with. It seems as if new numbers surface from time to time, including originals with undersized journals, which are not on this list. Materials can sometimes vary, so keep that in mind when trying to identify your Pontiac crank.

Year	Engine	Stroke (inches)	Main Journal (inches)	Crank Number	Material
1964–1965	389	3.75	3.00	9773383	Armasteel
1964–1966	421	4.00	3.25	9773384	Armasteel
1966	389	3.75	3.00	9782646	Armasteel
1966	421	4.00	3.25	9782769	Armasteel
1967–1969	428	4.00	3.25	9782769	Armasteel
1967–1970	400	3.75	3.00	9795480	Nodular Iron
1968–1970	350	3.75	3.00	9793573	Nodular Iron
1968–1970	400 Ram Air	3.75	3.00	9794054	Armasteel
1970–1974	455	4.21	3.25	9799103	Nodular Iron
1971–1974	350	3.75	3.00	481379	Nodular Iron
1971–1974	400	3.75	3.00	481380	Nodular Iron
1973–1974	SD-455	4.21	3.25	495030	Nodular Iron
1975	400	3.75	3.00	496414	Precision Cast
1975–1976	455	4.21	3.25	496453	Nodular Iron
1976–1978	350	3.75	3.00	496452	Precision Cast
1976–1979	400	3.75	3.00	499864	Precision Cast

Armasteel is a specific iron alloy developed by GM that possesses some steel-like strength qualities. Pontiac used it when casting certain rear axle housings, connecting rods, and crankshafts. Armasteel cranks were commonly used during the 1960s for specific performance applications where additional crank durability was required. Such cranks can be identified by the Armasteel name cast into it, and they make an excellent choice for performance use.

Selecting a Stock Crank

If considering an original Pontiac crankshaft, I suggest any Armasteel or nodular iron casting produced from 1964 to about 1974, so long as it contains

Cranks cast of nodular iron are likely the most popular Pontiac units available today. Pontiac used the alloy during the late 1960s and through the mid 1970s; such cranks have a large "N" cast into them. Some nodular cranks of the era lack the identifier and that can cause confusion. Pontiac records indicate that the cranks were still constructed of nodular iron, however.

Generally speaking, all Pontiac cranks were machined to accept a pilot bushing during production, but one occasionally surfaces that wasn't machined accordingly. While it's of no concern for an automatic transmission, it can be a significant issue for a manual transmission, and is one that's not found until the assembled engine is reunited with the transmission. Correction requires complete engine disassembly. I suggest checking that your crankshaft is machined for a pilot bushing before sending it to the machine shop. It's an easy fix that doesn't limit transmission choices down the road.

Before machining a crankshaft, any quality shop verifies that it's crack free with a magnetic particle inspection. Once deemed usable its main and rod journals can be machined to the appropriate undersized dimension. Since aftermarket cranks for all makes are such an affordable option, many machine shops are not investing in crankshaft refinishing equipment. Older shops generally still have it and can prepare your crank accordingly.

the appropriate stroke and journal dimensions for your application. These castings generally weigh between 60 to 70 pounds and are adequate up to 600 hp. Some push that toward 700 hp or slightly more, but it requires proper preparation and machining and precise tuning to survive reliably at that level.

Castings produced prior to 1964 are about as durable as later castings, but snout length is generally shorter, and the flywheel flange bolt pattern is different. That can present installation issues and/or require other corresponding parts such as the timing cover or flywheel/flexplate. The block may require minor grinding to achieve sufficient clearance when combining certain cranks and blocks. The crank counterweights or rod throws can contact the bottom of the cylinder bore at certain points during rotation. I highly recommend test fitting the components anytime a block and crank are not matched originals.

In an attempt to shed engine weight and improve fuel economy during the mid 1970s, Pontiac lightened its V-8 crankshaft of the era as well. The casting isn't quite as robust as earlier examples.

Such castings can be identified by the part number, which can be fully cast, partially cast and partially stamped, or fully stamped on the front counterweights. While completely adequate for engines producing up to 500 hp or so, an earlier Armasteel or nodular iron crank is a better option when pushing horsepower beyond that.

The highly specialized 1960s Super Duty and tunnel-port Ram Air engines were equipped with high-quality, forged-steel crankshafts for maximum durability and reliability. The factory forgings can be identified by a very wide parting line that runs parallel to the crank centerline, and further identified by its part number. Production was very limited. They remain extremely rare and are quite valuable. Unless you have an original engine that requires such a crank, I recommend leaving them to the restorers. Modern forgings are a better value.

Original Crank Modifications

A crankshaft constantly flexes during normal operation. Over time, cracks can develop in the fillet area (where the journals meet the counterweights). If your build includes a cast Pontiac crankshaft, it must be magnetically inspected (Magnafluxed) to determine that it's completely crack free before spending any money on machining. You're much better off spending a few bucks to ensure a crank you already purchased is usable and throwing it away than chancing the entire engine should it fail.

Pontiac used excellent materials and precision casting and machining techniques when producing its cast cranks. That and the use of quality bearings negated the need to artificially harden the journal surfaces to improve durability. It also eliminated the risk of removing that hardening when undersizing the journals

Micro-polishing leaves behind a very smooth journal surface that improves lubrication and bearing life. It involves a special polisher that circulates fine-grit polishing ribbons. It should be considered part of any rebuild, no matter how basic.

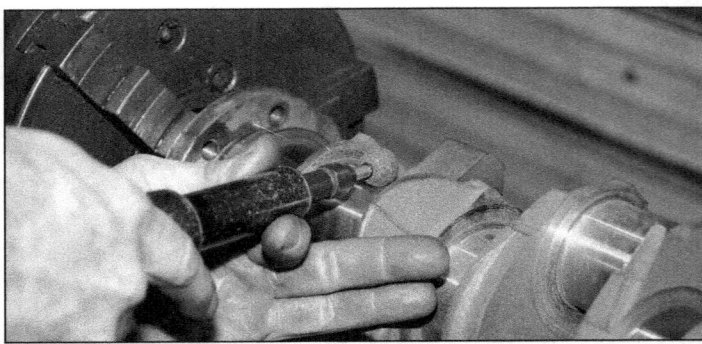
Chamfering the oil passages on a stock Pontiac crankshaft improves lubrication by giving the passage a greater surface to disperse oil. It removes burrs that can scratch the crankshaft and/or inhibit proper oil flow. It's a modification that your machinist should perform before journal polishing.

A grinding stone smooths the rough parting lines left behind on the crankshaft after the production process. It reduces the risk of cracks forming along the sharp edges, which can lead to failure in extreme applications. It's a simple process that takes just a few minutes to perform.

The oil passage in the main journal provides the rod journals with pressurized oiling. Most original cranks had only one oil hole and that can shut off connecting rod oil supply for a limited time during crankshaft rotation. Pontiac drilled the passage completely through the journal on certain 3-inch cranks and all 3.25-inch units. This "cross drilling" provides additional lubrication, and improves bearing life in high-revving engines with certain bearings. While most aftermarket cranks are cross-drilled, it's not necessary with main bearings that are grooved 3/4–inch around the surface.

during routine machining. About as much as .050 inch can be removed without compromising crankshaft rigidity. The limiting factor is availability of correctly undersized bearings.

Beyond normal grinding and micro-polishing of the journal surfaces, there isn't much more that a cast Pontiac crank requires for use up to 600 hp, or slightly more. I suggest removing all traces of casting flash from the counterweights to eliminate areas where cracks can propagate. Other modifications such as lightening or narrowing the counterweight's leading edge can increase engine acceleration rate, but may not improve actual performance in all instances and, as such, is not always cost effective.

The main journal oiling passages on a stock 3.25-inch Pontiac crank are drilled completely through for maximum rod-bearing lubrication. Those on a 3-inch crank are drilled only partially through. Drilling completely through the journal of a 3-inch crank can increase the amount of oil connecting rods see during each complete rotation. While there aren't any real negatives to cross drilling a stock crank, it isn't always required with modern main bearings. Chamfering the lubrication passages that intersect the journal surfaces removes burrs and better disperses oil over a greater area of the bearing surface.

If the journals of a stock cast crank require resizing to a dimension that exceeds modern bearing undersize availability, some specialty shops claim the crank can be salvaged by restoring the

The Pontiac crankshaft design allows a significant amount of main and rod journal overlap, which heavily contributes to its rigidity. It also allows the rod journal to be undersized and its stroke lengthened without grossly affecting its reliability. Adding stroke length increases displacement while improving torque output at low to moderate RPM.

The stock 455 block can tolerate a stroke length of as much as 4.25 inches or slightly more without issue. Whether you choose to modify an original casting or purchase an aftermarket cast or forged unit, expect it to drop into place without modification. It's an easy upgrade.

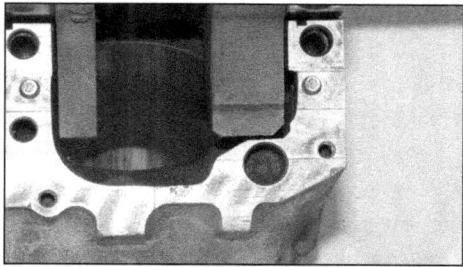
When combining a stroker crankshaft with a 400 block, there may be a few areas that require grinding to gain sufficient rotational clearance for the counterweights. Test fit your crank and rotate it through several times to ensure that the counterweights and rod throws do not contact the block at any point. This generally isn't an issue with a 455 block, but it cannot hurt to check.

thickness of the journal surface with spray welding. I strongly recommend against this for any high-performance engine, as any attempt to weld new material onto an existing cast surface can ultimately lead to cracking. If appropriately undersized bearings are unavailable, it's best to simply start with another crankshaft.

Pontiac did not use nitride or cryogenic treatments or hard chroming to harden its crankshaft journal surfaces. Such treatments, however, can benefit modern applications where extreme conditions and harder bearings can damage the journals. While such treatments may possess no negative effects in any rebuild, it certainly doesn't fit every budget.

Stroker Cranks

Adding displacement is an easy and effective way to improve engine output. In addition to enlarging bore diameter over the years, Pontiac often lengthened V-8 crankshaft stroke to accomplish exactly that. The additional stroke length creates more leverage on the crankshaft, which generally increases the amount of available torque, particularly at relatively low RPM, making it ideal for engines that are primarily street driven.

If you've ever driven a vintage Pontiac with a stock 400 and another with a stock 455, you likely recognized that the 455 felt more powerful, especially in its ability to spin the tires from a standing start. Though the two engines have nearly similar bore diameters at 4.12 and 4.15 inches, respectively, the additional torque the 455 generates is directly related to crankshaft stroke length, which is nearly 1/2 inch longer than that of the 400. Whether or not the 455 creates more high-RPM horsepower than the 400 depends upon many other factors.

The generous amount of journal overlap that Pontiac designed into its crankshaft allows a slight stroke length increase without grossly affecting the rigidity of a stock casting. The "stroking" process is quite involved and requires resizing the rod journals from the stock Pontiac diameter of 2.249 to 2.2 inches, a measurement that's common to the big-block Chevy. The rod journal axis is then relocated with offset-machining, which effectively increases stroke lengthen by .040 to .050 inch, adding another 6 to 8 ci of total engine displacement.

In addition to a slight displacement boost, greater stroke length increases the amount of time that a piston remains stationary at the top and bottom of the cylinder as the crankshaft changes direction. That "dwell" allows the piston to accelerate slower from a stop, which gives the engine more time to better fill and evacuate the cylinders. It also gives the pressure generated during combustion more time to exert its force on the piston, which promotes maximum torque. It can, however, induce engine-damaging detonation if the engine is already running on the edge for the fuel octane being used.

Extended piston dwell does present some negatives. Even though the piston accelerates slower, it actually reaches a higher terminal speed as it has a greater distance to travel during each rotation. That and the additional side loading associated with the longer stroke translate into greater friction and stress exerted on the block and its cylinder walls. A longer-than-stock connecting rod (to improve rod-to-stroke ratio) can alleviate some of that, but a rod that's too long requires a piston with a very short compression height, and that can lead to its own set of issues.

With a number of new crankshafts being produced for the Pontiac V-8 in recent years, manufacturers began offering Pontiac cranks with a wide array of journal and stroke combinations. That includes stock dimensions for use as new OE replacements, and specialized units that feature stock journal main sizes but with much greater stroke lengths. The advent of the readymade "stroker crankshaft" allows any hobbyist to easily transform a 400 into a 461- to 467-inch engine at a very reasonable price, and it's an immensely popular modification.

Aftermarket Cranks

Eagle Specialty Products introduced a new Pontiac V-8 crankshaft casting in 2001. It provided an alternative to using a questionable original, or where journal undersize requirements exceeded available bearing options. Other companies have introduced similar castings in recent years, and while some are of questionable quality, I am comfortable recommending

Ohio Crankshaft offers new Pontiac crankshafts constructed of cast iron or forged steel in a variety of popular bore and stroke dimensions for blocks with 3- and 3.25-inch main journals. Cast units like this are constructed of nodular iron alloy and make an excellent choice when replacing an original piece. Its forged cranks are ideal for high-performance applications producing 1,000 to 1,200 hp.

Rear Main Seal Lubrication Grooves

The rear main seal surface of a Pontiac crankshaft contains small grooves that lubricate the original rope-type seal during normal operation. Without sufficient lubrication, the rope seal can shrivel and/or burn up due to friction. It almost always results in seal failure of some type, and often an oil leak that requires major engine disassembly to properly repair. While present on all original Pontiac cranks, these serrated lubrication grooves are not all present on aftermarket cranks.

With an aftermarket lip-type rear main seal, the grooves can tear the lip surface as the crankshaft rotates, potentially leading to oil leaks. The grooves on a well-used original crank are usually sufficiently worn and potential seal damage isn't an issue. Those on a new aftermarket crank sometimes are sharp, however, and if the seal surface is rough to the touch, your machinist can polish it like any other crankshaft journal to smooth it. BOP Engineering recommends removing no more than .006 inch from the journal diameter for maximum compatibility with its lip-type seal.

No modifications are required when combining a lip-type seal with an aftermarket crankshaft that lacks seal surface grooves. The absence of lubrication grooves on an aftermarket crank can prevent the use of any woven seal such as that by Best Gasket, however. As with the lip seal, rough grooving on an aftermarket crank with serrations can shred a rope seal in short order too, so the seal surface may still require polishing with a rope seal as well. Your machinist can help you decide if your particular crankshaft is compatible with your rear main seal choice.

The rear main seal surface of a factory Pontiac crankshaft contains small grooves that lubricate the original rope seal during normal operation. Most aftermarket cranks contain these same grooves, but some do not. If the grooves are rough to the touch, they may require polishing to avoid seal damage. The seal surface of those without the grooves is polished like any other journal. If your crankshaft lacks lubrication grooves, it isn't compatible with a rope-type seal.

cast crankshafts from a few of them.

When the carbon content of iron reaches 6 percent, an alloy becomes steel in a technical sense. For its cast cranks, Eagle uses nodular iron with a high carbon content that can be classified as steel. Eagle's cranks are produced at its own China-based production facility and are held to a strict tolerance. While some isolated thrust surface issues were reported early on, today the Eagle casting is known for its consistency and reliability. I consider it an excellent choice whenever a stock nodular iron crankshaft is considered.

Weighing about 70 pounds, Eagle's cast crank is available with 3-inch main journals and 4.25-inch stroke. A 3.25-inch main journal unit is available with stroke lengths of 4.21 and 4.25 inches. The 4.21-inch unit uses stock-type connecting rods while the 4.25-inch units use longer big-block Chevy type of rods, such as those measuring 6.7 and 6.8 inches. Eagle's recommended horsepower limit for its cast Pontiac crank is 700 hp.

Ohio Crankshaft began offering cast-iron Pontiac V-8 crankshafts within the past several years. The castings are constructed of nodular iron and weigh about 70 pounds. I have found excellent fit and function. Cast cranks with stock Pontiac rod journals are available with 3- and 3.25-inch main journals, and with 4- and 4.21-inch stroke lengths. A stroke length of 4.25 inches is only available with 2.2-inch rod journals. A cast crank from Ohio sells for less than $300 and can be considered another excellent choice for engines producing as much as 600 hp.

When an engine exceeds 550 hp or 6,000 rpm, or when forced induction or nitrous oxide is used, a forged crank is a very worthwhile investment. A cast crank can flex at very high RPM and/or in high-horsepower applications, causing premature bearing wear, especially when using a unit with 3-inch main journals, and a stroke length of 4.21 inches or more.

Eagle's cast and forged-steel crankshafts are very popular. They offer the quality and reliability required for high-performance applications. The forged cranks, which are constructed of 4340-steel alloy, are considered by many to be among the very best available today.

A forged-steel crankshaft is generally more resilient to flexibility and fatigue and can sustain 1,000 to 1,200 hp for only a few hundred dollars more than the cost of a cast crank. Even if you don't expect to ever reach that performance level, a forging allows for significant power increases in the future without creating reliability concerns. Replacing a cast crank with a forged unit in an otherwise good-running engine can be a costly and time-consuming endeavor! There are a few excellent choices readily available that I am confident to recommend.

Eagle introduced its forged 4340-alloy crankshaft in 2009. Selling for around $800, it's an affordable option that offers the durability required for high-performance use. A fully-machined forging weighs about 75 pounds and includes cross-drilled main journals for maximum oiling. For blocks with 3-inch main journals, Eagle offers stroke lengths ranging from 4.21 to 4.5 inches. The 4.21-inch unit features 2.25-inch rod journals while others have 2.2-inch journals. Forgings with 3.25-inch main journals are available with 4.21- and 4.25-inch stroke lengths only.

Forged 4340-steel cranks from Ohio Crankshaft are available with 3- and 3.25-inch main journals. The 3-inch unit

Most aftermarket cast and forged Pontiac cranks are available with a stroke length up to 4.25 inches. Some companies offer a forging with a stroke length of 4.5 inches. The extra .25 inch can greatly stress the cylinder walls of a stock block, however. While installing a 4.5-inch crank into a stock block is certainly possible, an aftermarket block is much better suited for it.

features cross-drilled main journals and are available in a variety of stroke lengths ranging from 3.75 to 4.75 inches with Pontiac- and Chevy-size rod journals. The 3.25-inch forgings with stock-dimension rod journals are available with 4- and 4.21-inch stroke lengths, and 4.25- and 4.5-inch stroke lengths with 2.2-inch journals. Weighing around 75 pounds, Ohio Crankshaft claims its forging can sustain as much as 1,500 hp. With a cost around $600, it's an excellent value.

Scat Crankshafts offers a variety of forged 4340-steel crankshafts with 3- and 3.25-inch main journals in stroke lengths of 4, 4.25, and 4.5 inches. Its cranks are only available with 2.2-inch rod journals and come in three different weight variations, which are accomplished with counterweight profiling. Its Standard crank weighs about 75 pounds, while the Lightweight and Super Lightweight cranks weigh about 65 and 60 pounds, respectively. Costs range from $850 for a Standard unit to $1,525 for a Super Lightweight forging.

RPM Industries and Star Galaxy also offer cast-iron and forged-steel crankshafts for the Pontiac V-8. A variety of main and rod journal and stroke length combinations are available. I understand the quality is generally good, but must admit that I have no direct experience with examples from either company. That doesn't suggest that the offerings are inferior, however. I simply haven't had the opportunity to work with them. Your Pontiac vendor should be able to comment on quality and pricing.

Many crankshaft companies also offer forged-steel connecting rods. Several have created complete rotating assembly kits that include a new cast or forged-steel crankshaft and forged-steel connecting rods and combine it with forged-aluminum pistons and the required rings and bearings to create a complete rotating assembly package. Most often the crankshaft features a stroke length of 4.25 inches, and it's combined with connecting rods that measure 6.7 to 6.8 inches, creating a "stroker" assembly.

A stroker package is an easy way to significantly increase the displacement and output of 350- and 400-inch engines with pricing that starts at about $1,500. A typical 400 block that's been bored .060 inch can quickly and easily displace as much as 467 ci simply by using an aftermarket crankshaft with a 4.25-inch stroke. A 455 can be made as large as 474 ci using similar equipment. Longer strokes are available and can be used, but it sometimes requires special pistons and additional block preparation such as clearance grinding, block filler, and four-bolt main caps. Your machinist or Pontiac vendor can help you decide which components are required to help you achieve your performance goals.

Specialty Billet Cranks

A crankshaft machined from a billet of 4340-steel alloy is another very durable option that's popular with Pontiac racers. The argument of whether a billet crank is actually more durable than a forged crankshaft has raged on for many years. Ask any professional engine builder or crankshaft manufacturer and you'll hear some very strongly supported arguments and opinions. You can decide if one is required for your application; here, I provide some comments about them that may better educate you about billet crankshafts in general.

The grain structure in a billet of high-quality steel is very dense and runs in a similar plane. As the material is heat treated to a specific temper it alters the grain structure and improves strength, which further improves rigidity, reducing the amount it flexes during normal operation.

In my opinion, the main advantage of a billet-steel crankshaft is the ability to create a custom unit that fits a very specific application. A company such as Moldex can produce a very high quality crankshaft that can sustain about any amount of horsepower or RPM possible by tailoring counterweight design and other aspects for the intended application. Any combination of rod and main journal dimensions are available, as are popular features such as knife-edged counterweights and gun-drilled journals

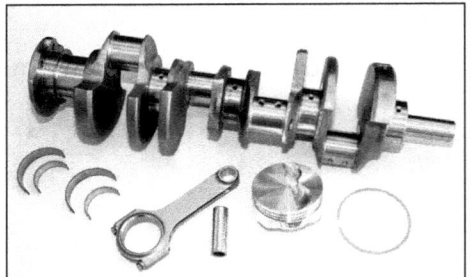

Most aftermarket crankshaft manufacturers and professional Pontiac engine builders offer complete rotating assembly kits. They includes a cast or forged crankshaft with stock-length or longer stroke, forged connecting rods and pistons, and the appropriate rings and bearings. Such kits typically require nothing more than simple block machining for drop-in installation. They are an excellent performance value that can quickly and easily improve the performance and reliability of any engine build.

Balancing the rotating assembly provides smooth and consistent engine operation and improves bearing life. It's something I recommend for any engine build and should be a service your machine shop is well equipped to provide. When purchasing a complete rotating assembly kit, some sellers offer a balancing option at extra cost.

to reduce weight. Pricing from Moldex starts at $3,250 and because each crankshaft is produced on a custom-order basis, you need to allow for a lead time of 12 weeks or more.

Balancing

Balancing is an important part of any engine build. It's one that any quality machine shop should consider standard practice during a rebuild, no matter how basic. It consists of weight matching the components of the entire reciprocating assembly to a very close tolerance. While some rotating assembly kits that include a new aftermarket crankshaft are balanced by the Pontiac vendor prior to shipping, your machinist can perform the task at a reasonable cost if using outside components.

For any max-performance build, balancing should include weight matching the connecting rods, wrist pins, pistons, and even the rings. A fixture replicating the mass of those components is then bolted to the crankshaft journals, and the assembly is spun at a relatively low speed. Material is then added or removed from the crankshaft counterweights until the complete assembly is balanced accordingly. The harmonic damper and flywheel/flexplate are often installed at some point during the process to ensure neither alters the balance unexpectedly.

Contrary to what many believe, balancing doesn't necessarily improve engine output. It promotes maximum performance while reducing harmonic vibrations at certain engine speeds. That provides smooth and consistent engine operation at all speeds, and that can reduce complete engine wear over its lifetime, particularly to the bearings.

Harmonic Dampers

As a crankshaft rotates during normal operation, some pistons are forced downward by combustion, while others are pulled downward or pushed upward. Occurring hundreds of times per second, it causes the crankshaft to flex or twist, and that creates torsional vibration, which can eventually cause the crankshaft to fatigue and fail. Fastened to the front of a typical crankshaft is a hub assembly that's designed to counteract, or absorb, the harmonic irregularities to

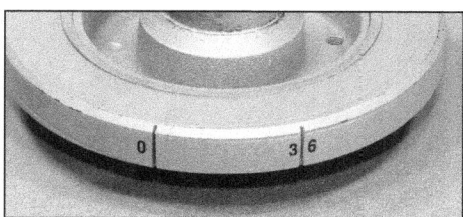

The outer ring of the stock Pontiac damper is marked for top dead center (TDC) only. This makes total timing adjustment an impossible process unless you own a dial-back timing light. A popular modification is to add one or more marks that represent other timing points. Indexing is easily accomplished by measuring the circumference, dividing it by 360 degrees, and multiplying by the desired timing number. Simply scribe another mark that distance from the TDC mark in a clockwise manner. I routinely label mine for easy visual identification.

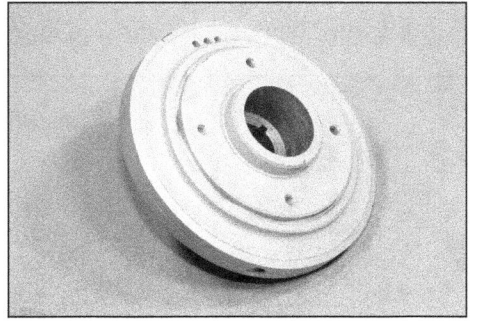

The stock Pontiac harmonic damper introduced in 1968 is a pressed-together design that's most common. It worked quite well in original applications, but it's not uncommon to find the rubber isolator that separates the hub from the outer ring has lost its dampening ability from more than 30 years of use. NOS units are still available and are suitable for stock rebuilds, but I recommend an aftermarket unit for high-performance use.

What is SFI Classification?

The SFI Foundation provides the automotive aftermarket and the motorsports industry with a system of developing and administering various standards, certifications, and testing criteria for chassis and engine components and safety equipment. Racing organizations adopt those standards as a way to improve the safety of its competitors.

The components found within a performance engine that most commonly receive SFI certification are the harmonic damper and flywheel/flexplate, but it can include many others. Not limited to competition engines only, SFI approved components can improve the reliability of any high-performance street or non-class race engine too.

Once a component or piece of safety equipment is certified, it receives an SFI certification label that must be visible or available for viewing during technical inspections. Those components must periodically be recertified, and that can include professionally rebuilding or refurbishing the existing assembly, or replacing it with a new one. More information about the SFI Foundation and the standards for the components it certifies can be found at www.sfifoundation.com.

effectively prevent crank failure over the life of an engine.

In addition to dampening harmonics, the assembly serves a second purpose for many engines of other makes. It can contain a slight imbalance, which must be factored in when balancing the entire reciprocating assembly, lending the name "harmonic balancer." Pontiac V-8s were balanced internally, however, and though "balancer" is often used to describe a Pontiac assembly in conversation, vintage Pontiac literature refers to it as a "harmonic damper." I will use "damper" for the sake of accuracy, but when considering a damper, it should be for an "internally balanced" engine, assuming your machinist balances your Pontiac in the traditional manner.

The inner hub slides onto the crankshaft snout. It's keyed for positive location and retained by a large bolt and washer assembly that must be torqued to 160 ft-lbs during installation. Original Pontiac and OE-type aftermarket dampers use an outer ring, or inertia weight, that floats about the hub. It's isolated by elastomer (rubber), which absorbs torsional irregularities, but some aftermarket units use silicone fluid. Either type can work quite well if appropriately weighted and sized for the application.

The outer perimeter of most aftermarket dampers is indexed with timing marks, which provides several advantages for assembly and tuning. It can make it easier to degree a camshaft or set and/or adjust initial or total spark lead.

Pontiac dampers produced before 1968 bolt together and can be difficult to work with. The original Pontiac damper

Powerbond balancers are an affordable option when considering a new Pontiac harmonic damper. Its entry-level OE replacement is ideal for engines producing 400 hp or more while its Race unit is SFI approved and ideal for engines producing 600 hp or more. Its dampers are indexed for easy timing adjustments and accept stock accessory pulleys.

BHJ's Pontiac damper is a high-quality piece constructed of billet materials. It installs easily, is indexed for easy timing adjustments, and accepts most stock accessory pulleys. It carries an SFI certification and is a solid choice for engines producing as much as 1,200 or more.

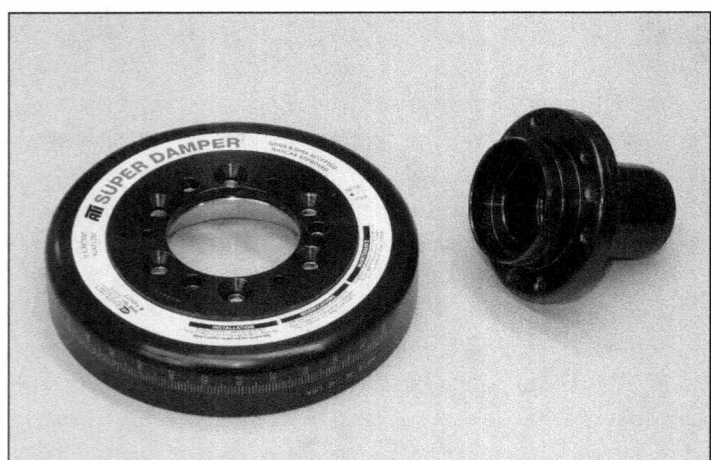

The Super Damper from ATI is a multi-piece design that bolts together and is SFI certified. Its outer shell is fully indexed and is available in steel for most street/strip applications or in aluminum for engines that accelerate very quickly. It's among the best damper options available today for max-performance engines.

The retaining bolt for a stock Pontiac damper calls for 160 ft-lbs of torque for proper installation. ARP offers new harmonic damper fasteners specifically for Pontiacs that require 150 ft-lbs of torque using its proprietary moly-based thread lubricant. ARP number-190-2501 is a typical bolt but number-190-2502 (shown here) accepts a 1/2-inch square-drive ratchet or breaker bar that allows for easy engine rotation.

introduced in 1968 is a unitized design that measures 6-inches in diameter. It's an excellent piece that performs very well in stock applications to those producing 400 hp. New units were available through Pontiac parts departments until a few years ago. Used units are at least 30 years old, however, and the rubber isolator can deteriorate and/or shrink, allowing the outer ring to rotate. That not only affects spark timing accuracy, it can affect the damper's balance, negatively impacting its ability to dampen harmonics.

In my opinion, after a thorough visual inspection and physically verifying that its outer ring hasn't slipped, a used 1968–1979 balancer is adequate for stock-type rebuilds or those with very mild performance modifications only. I strongly suggest one of the many aftermarket Pontiac dampers available today.

Powerbond offers three internally-balanced, stock-sized Pontiac dampers. That includes an OE replacement for engines producing up to 400 hp, a Street Performance unit for use up to 600 hp, and an SFI-certified Race Performance damper for applications beyond that. The OE replacement and Street Performance units use a cast hub and outer ring and sell for around $100. The Race Performance unit uses a forged hub and outer ring and sells for less than $200. Powerbond injects rubber onto the hub and ring assemblies during production to positively secure the components for maximum strength and dampening ability. While Powerbond dampers cannot be rebuilt, the reasonable cost makes replacement a cost-efficient solution.

BHJ produces quality harmonic dampers for many engines including the Pontiac V-8. Its damper features a billet-steel inertia weight that's isolated by rubber from a crankshaft hub constructed of billet steel or aluminum. The steel hub is available with a stock-type slip fit, or a press fit for high-performance applications. BHJ's damper is SFI certified, measures 6.8 inches, and is internally balanced. While designed to accommodate other factory accessory pulleys, it is not compatible with the original A/C units. The BHJ balancer sells for about $450 and is ideal for engines producing as much as 1,200 hp, or even more.

ATI's SFI-certified Super Damper is a multi-piece design with a crankshaft hub constructed of high-quality steel. The internal inertia weight is isolated by rubber and encased by inner and outer shells available in steel or aluminum, and in two different diameters. ATI recommends its 6.325-inch damper for engines producing as much as 600 hp and its 7-inch unit for applications beyond that. The Super Damper, which sells for about $400, is internally balanced and fully rebuildable. It does not always accept stock pulleys, however. In my opinion, the Super Damper is among the very best for max-performance applications.

Engine Bearings

Engine bearings are used to support components within the engine that are in constant motion, such as the crankshaft and camshaft. A hydrodynamic oil wedge prevents the component journals from actually contacting the bearings during normal operation. A bearing must be durable enough to prevent distortion or deformation under myriad loads for extended periods, yet soft enough that it allows dirt and foreign material to imbed into it as opposed to damaging the journals. It must also resist corrosion and heat.

Considered wear components, bearings are designed to survive an engine's lifetime, assuming it's operated within its intended parameters and regularly maintained. That duration may be 100,000 miles or more for passenger car service or as much as several dozen passes in a high-end race engine. At the time of rebuild or refresh, a normal running engine with properly selected bearings leaves the component journals in the best possible condition, requiring the least amount of preparation for use with new bearings.

Modern bearings feature a multi-layered construction that consists of an aluminum or steel shell (or backing) for support, an intermediate layer for strength, and a soft, outer overlay. That overlay is generally comprised of a soft metal, such as a lead alloy, which offers a high level of "imbedability" to absorb dirt and particles without damaging the crankshaft journals. Babbitt is another very soft, lead-like material that's also commonly used. Neither is capable of enduring high loads for long periods without flaking apart if layered too thick, however.

Combining a thin overlay of lead or Babbitt with an intermediate layer of copper, lead, and/or nickel allows the bearing to perform its intended function for extended periods. Such tri-layer or multi-layer bearings are commonplace in today's industry. While Babbitt is still used in some instances, an overlay of lead, tin, aluminum, or some combination of them is often used in specialty bearing sets where heavier loads are common.

Crankshaft bearings are a two-piece design where the upper half resides in the block or connecting rod, and the lower half resides in the corresponding cap. ACL, Federal-Mogul, King, and Mahle/Clevite (Clevite) are among the best crankshaft OE replacement-bearing suppliers for Pontiacs today. Depending upon the manufacturer, bearings are available in standard dimensions, .001-inch undersize and then in .010-inch increments. Present undersize is up to .030 inch for main journals and .040 inch for Pontiac- and Chevy-size rod journals. You may need to check with the manufacturer for current availability.

CHAPTER 3

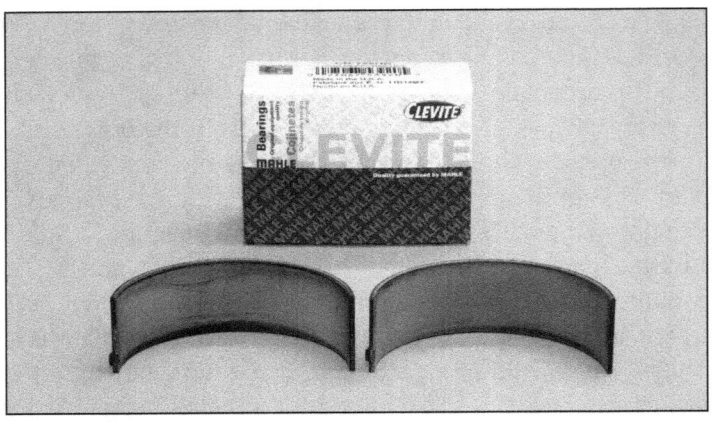

Connecting rod bearings are solid bearings that carry a significant amount of load. The bearing must be durable enough to not distort or deform under extreme pressure, but must be soft enough that dirt and metallic particles imbed into its surface without scoring the crankshaft. Modern bearings offer multi-layer construction, which provides the best operation compromise. Clevite rod bearings are among the most popular with professionals.

Main journal bearings are under continuous pressure. Modern performance sets include an upper bearing that's fully grooved to provide the rod journal with pressurized oiling. Low-performance sets typically include fully grooved lower bearings, which sacrifices precious surface area by compromising its durability and load-carrying capability. When combined with a cross-drilled crankshaft, a main bearing set like this provides excellent durability.

Clevite and Federal-Mogul bearings seem to be the most popular with Pontiac engine building professionals. In addition to OE lines, either company also offers a "performance" bearing line designed for higher loads. Performance bearings are generally harder, and while better at enduring heat and heavy loads, they can resist some materials that can otherwise damage or score the crankshaft journal surfaces. Performance bearings can also be tougher on the journals of a cast crank. An artificial surface hardening process is a worthwhile consideration when using specific bearings in certain applications.

Main Bearing Grooves

All Pontiac main bearings are grooved to provide a path for pressurized oil to access the rod journal lubrication passage. While it might seem that using a main bearing set with upper and lower halves that are fully grooved would provide maximum rod lubrication, the grooving reduces bearing surface area, and that can compromise its load-carrying capability. While some main bearing sets are available with full grooves in each bearing half, and they may be completely adequate for a stock-type rebuild, I don't consider them the best choice for very high performance use. Failure can occur at the highest loaded area, which is found at the lowest point in the main cap.

A crankshaft with cross-drilled main journals, such as the Pontiac 455, can maintain constant rod journal oiling while supporting the greatest load by using a main bearing set where the upper half is fully grooved and the lower half is solid. Most factory 3-inch crankshaft castings aren't cross-drilled, however. For many years, cross-drilling the main journal and using a half-grooved main bearing set was the only way to attain maximum rod journal lubrication and load-carrying capability in such combinations. That's no longer the norm.

Within the past several years manufacturers have developed main bearing sets with a fully-grooved upper half and partially-grooved lower half. In addition to constantly supplying oil to the connecting rods, the partially-grooved bearing provides full bearing width at the highest loaded area and places pressur-

Federal-Mogul offers main bearing sets for the Pontiac V-8 that include a fully grooved upper bearing and partially grooved lower bearing. The intent is to provide maximum connecting rod oiling without compromising bearing durability. The partially grooved lower bearing also places pressurized oil near the heaviest loaded point, and that improves bearing performance in extreme applications.

ized oil very close to it. Grooved across about 220 degrees of the journal surface, these "3/4 grooved" main bearings are the choice of most Pontiac engine builders for their excellent lubrication and surface

CRANKSHAFTS

Bearing clearance is most accurately measured using a dial bore gauge with recordings taken 90 degrees from the bearing parting line. Subtracting that measurement from the crankshaft diameter leaves you with the exact amount of clearance. Main and rod bearing clearance should each measure at least .0025 inch, depending upon the application. Similar results can be attained using Plastigage or a similar product if a dial bore gauge isn't available.

Crankshaft thrust is taken up on the number-4 main journal. Whether using an original Pontiac crank or one of the main aftermarket options, thrust clearance should remain between .005 and .009 inch. Any variance outside that range can lead to premature bearing wear and/or abnormal operating issues. Selecting main bearings from another manufacturer can sometimes affect it, but I suggest contacting your machinist and component suppliers if it varies outside that range.

strength qualities. Engines with forced induction or nitrous oxide may require different bearings, so be sure to discuss it with your Pontiac engine building specialist.

Bearing Clearance

The excellent film strength of modern oil and state-of-the-art machining processes allow for adequate lubrication with bearing clearances that are slightly tighter than what you might find in older engines. Modern passenger car engines often specify 5W-20 or 5W-30 oil, which flows better through tighter clearances than heavier oil, while reducing horsepower loss and improving fuel economy. Lightweight oil isn't the best choice for all applications, however, particularly if the engine wasn't designed for use with it.

High-performance engines can generally tolerate slightly greater than normal clearance for many reasons. It increases oil to flow across the bearing surface at all times and allows the use of heavier oil such as 15W-40 and 20W-50, which doesn't thin out as much as lighter oil as it heats up. Circulating heavier oil consumes a few horsepower, but it can better protect bearing surfaces in myriad speeds and conditions. Some racers prefer lightweight oil to free up a few horsepower, but I feel it's best in instances where an engine is regularly torn down and the bearings can be inspected for lubrication-related wear, and adjustments made.

Generally, a high-performance engine requires about .001 to .0015 inch of bearing clearance for each inch of journal diameter. Pontiac originally specified about .002 inch of rod journal clearance and .0025 inch of main journal clearance for its V-8s. I consider that an excellent starting point for most high-performance rebuilds, but your engine can likely tolerate closer to .003 inch or slightly more without issue. It's difficult to suggest a specific amount of bearing clearance that applies to every application, however, because it can vary with such factors as operating range, expected load, cylinder pressure, oil pressure, and oil viscosity.

It should also be noted that not all bearings are created equal. Some are tighter than others. Federal-Mogul bearings generally run about .0005 inch more clearance than others. For example, if you find your clearance a bit tight with Clevite, a set of Federal-Mogul bearings might provide you with the amount of the clearance you need. Other bearings, such as those from Clevite with an "X" in the part number, are designed to provide .001-inch greater clearance than a standard bearing. Clevite's H-series bearings that contain an "N" in the part number are narrowed slightly to provide greater crankshaft fillet clearance when using aftermarket units. Federal-Mogul's performance bearing shares a similar characteristic.

CHAPTER 4

CONNECTING RODS

The most common Pontiac connecting rod is the cast unit constructed of Armasteel, a specific iron alloy that boasts some steel-like qualities. While suitable for a mild-performance rebuild, entry-level, stock replacement forgings are a much better option, particularly anytime performance is increased beyond stock levels.

Pontiac developed a beautiful forged-steel connecting rod for its Super Duty 455. The high-quality forging is constructed of premium steel, uses 7/16-inch fasteners, and is heat treated for maximum durability. Very expensive when new, SD-455 rods remain quite valuable and are suitable for high-performance use. Modern 4340-steel forgings may be a more affordable option, however.

A connecting rod may be the most highly stressed component in an engine. Essentially in constant motion during normal operation, it's designed to endure compressive loads and change direction nearly instantaneously, and the inertial force upon it increases with engine speed. A connecting rod for a particular application must be constructed of a sturdy material that's elastic enough to not fatigue. There are two types commonly associated with Pontiacs: cast and forged.

Original Pontiac Rods

Cast connecting rods are very cheap to produce. Cast iron is generally very strong up to its elastic point, but tends to shatter once surpassed. A properly designed cast connecting rod can provide a long service life if an engine is operated within its intended limit, making it ideal for production engines. Steel forgings are durable but somewhat expensive to produce. Heat treating to a specific hardness is required to survive the greater inertial forces associated with power increase and/or high engine speeds, making it even more costly.

CONNECTING RODS

How Much Do Pontiac Connecting Rods Weigh?

Connecting rods are just one set of components that creates the rotating assembly. Like the pistons and crankshaft, connecting rods must be of appropriate weight and design to allow for proper engine balancing, and be of sufficient durability for the intended acceleration rate and operating rate of the particular engine being built. This chart contains the approximate weight of many stock-type connecting rods that are popularly used during Pontiac V-8 rebuilds.

Manufacturer	Description	Alloy	Approximate Weight (grams)
Pontiac	Stock cast I-beam	Cast Armasteel	900
Pontiac	Super Duty 455 I-beam	8610-steel	850
RPM International	Stock replacement I-beam	5140-steel	750
RPM International	Maxx Lite I-beam	4340-steel	700
RPM International	Pontiac I-beam	4340-steel	725
RPM International	Pontiac H-beam	4340-steel	760
Eagle	Pontiac H-beam	4340-steel	760
Crower	Pontiac I-beam	4340-steel	850
CP-Carrillo	Pontiac H-beam	4330-steel	830
Oliver	Pontiac I-beam	4340-steel	820
GRP	Pontiac I-beam	Aluminum	640

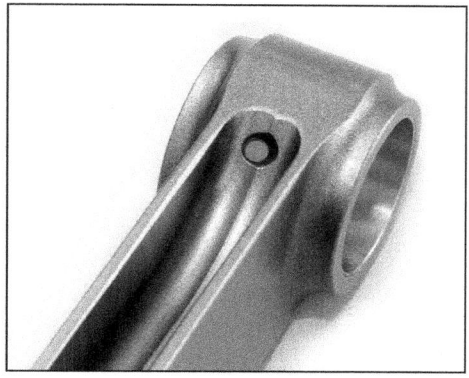

Many connecting rod manufacturers offer a variety of options aimed at improving reliability in specialized applications. That includes this hole that provides additional wrist pin oiling via crankcase splash. Another popular option is a passage drilled through the beam that provides a direct supply of pressurized oil to the wrist pin for applications with extreme cylinder pressure.

Depending upon the year and/or engine application, Pontiac used cast or forged I-beam connecting rods. A low-grade forging with minimal heat treating was used early on, and it wasn't dimensionally stable. During the 1960s Pontiac introduced a strong cast unit constructed of Armasteel, a GM trade name for a pearlitic malleable iron alloy that possesses some steel-like qualities. Because of age, stability, and durability concerns, I strongly feel that neither of these original Pontiac rods should be used in any high-performance engine.

Pontiac used high-quality steel forgings in its max-performance efforts such as the Super Duty 421 and Ram Air V, and they are extremely rare and difficult to find today. The 1973–1974 Super Duty 455 used a completely new forged-steel connecting rod for maximum durability. The specialized forging and finishing process was outsourced to Teledyne Continental Motors in Michigan. Orders from racers for the premium stock-length rod poured into the Parts Department, and at one point Pontiac reported more than 1,300 orders at one time.

Once Pontiac discontinued the SD-455 after the 1974 model year, its forged-steel rod was produced and remained available through Pontiac's Parts Department on a limited basis for several years. The rods were so rare and costly that many couldn't afford them, however. While new old stock (NOS) and used SD-455 rods appear for sale from time to time and are certainly suitable for a modern high-performance build, they remain very valuable. For the total cost involved I believe there are more affordable options on the aftermarket.

Aftermarket Steel Rods

Eagle Specialty Products gave Pontiac racers the durability they desired but couldn't otherwise afford when it introduced its high-quality 4340-steel forging in the mid 1990s. A great number of forged-steel connecting rods for Pontiacs followed and there is a wide variety available on the aftermarket today. Be sure to discuss your build plans with your Pontiac engine builder or parts vendor to determine which connecting rod is best suited for you.

RPM International

RPM International offers a stock-replacement Pontiac connecting rod constructed of 5140-steel produced at its China-based facility. The new I-beam features stock Pontiac dimensions and 7/16-inch-diameter bolts. It's an excellent and affordable alternative to reusing cast originals in any performance engine generating up to 500 hp and/or a maximum engine speed up to 6,000 rpm. Available from most major

CHAPTER 4

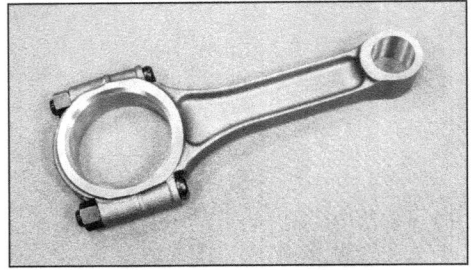

Within the past few years, RPM International introduced its stock replacement forging for Pontiacs. Constructed of 5140-steel, the I-beam features the same dimensions as a stock cast rod, but is rated up to 500 hp and 6,000 rpm. A complete set sells for about the same price as reconditioning eight cast stockers, making it a very popular performance value.

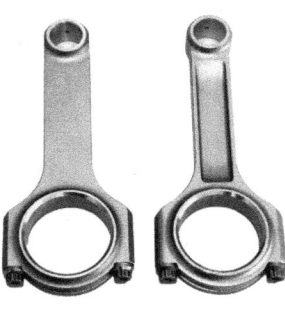

Along with its popular forged-5140 I-beam, RPM International also offers two 4340-steel connecting rods for engines producing up to 800 horsepower. Available in either H- or I-beam designs, and in stock dimensions or for stoker applications, the rods are an excellent choice if you're building a performance engine on a budget.

Pontiac vendors and marketed under various names, a complete set sells for less than $250.

Steel alloyed to 4340-spec offers excellent endurance qualities and is often used to produce high-strength connecting rods. I strongly recommend upgrading to forged-4340 rods in any engine producing more than 500 hp. Several companies produce forged-4340 rods specifically designed for the Pontiac V-8 applications, and RPM International is among them. Its 4340 I- and H-beam connecting rods are a popular option for engines producing up to 700 hp or slightly more. Its lightweight I-beam variant, called Maxx Lite A-beam offers similar durability.

I-Beam Versus H-beam

If connecting rods are so heavily stressed, it may seem that simply designing a larger, bulkier connecting rod would improve its performance and durability. That isn't the case, however.

A heavier connecting rod increases the overall weight of the reciprocating assembly, and that can make the engine rev lazily. In fact, a connecting rod that's too heavy can actually break the crankshaft. The best rod for any engine is one that's properly designed for the application load without being too heavy.

There are two distinct connecting rod designs commonly available for Pontiacs today. Generally speaking, an I-beam rod offers the greatest strength when a majority of compressive force occurs at an angle to rod beam. An H-beam offers greater columnar strength, or conditions when the compressive force is parallel to the beam. To accommodate the higher angular load, an H-beam is usually beefier, and that translates to heavier overall weight when compared to a similar-strength I-beam.

So which type is best for you? That largely depends upon the type of engine you're building. When considering the power level and engine RPM that most high-performance Pontiacs operate in, most high-quality aftermarket I- or H-beam rods are completely adequate, assuming that the chosen rod is compatible with the engine's intended power and/or RPM level.

An I- or H-beam may have district advantages over the other in a highly specialized application, such as a very high horsepower engine, very high operating speed, and/or when a power adder such as forced induction or nitrous is used. In those instances, it's best to discuss your options with your Pontiac engine building professional to determine which connecting rod is best for you.

Two types of connecting rods are commonly available for Pontiac V-8s: I-beam and H-beam. They are characterized by the cross-sectional shape of the main beam. Both types offer distinct advantages, and though either performs suitably in most performance rebuilds, sometimes one is better than the other in a specialized application. Your Pontiac parts supplier can help you determine that.

Eagle Specialty Products

Eagle's original forged-4340 H-beam remains very popular with Pontiac hobbyists. Eagle rods are finished to precise standards that result in better consistency and reliability. Selling for

CONNECTING RODS

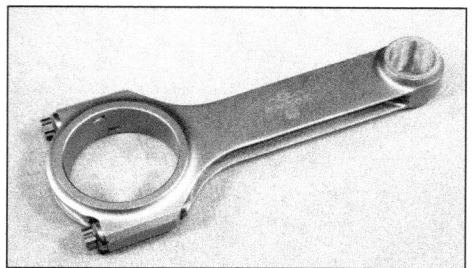

Eagle introduced one of the first affordable forged 4340-steel connecting rods for Pontiac V-8s during the 1990s. An excellent value and quite durable, the H-beam is very popular with performance enthusiasts. Available with stock Pontiac dimensions with pressed-fit or floating wrist pins, Eagle also offers a wide array of dimensions for stroker applications.

Crower supplies the hobby with many quality valvetrain components, but its Sportsman Big Block series connecting rods for Pontiac V-8s are considered among the best available today. Constructed of forged 4340-steel, the I-beam design offers excellent strength. They are available with pressed-fit and floating wrist pins and in popular stroker lengths too.

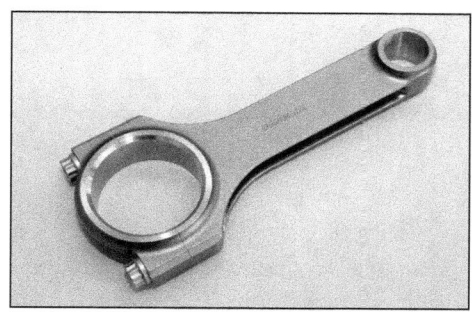

CP-Carrillo produces a number of premium forged-steel connecting rods that are compatible with the Pontiac V-8. Its proprietary H-beam design offers excellent durability for most severe applications. It also produces an NHRA-approved Pontiac rod, which features an overall length of 6.635 inches, but uses a 2.2-inch connecting rod journal and .990-inch-diameter floating wrist pin.

about $550 per set, it's an excellent choice for Pontiac engines producing up to about 750 hp and turning 7,000 rpm. An available fastener upgrade can increase that toward 1,500 hp.

Crower Cams and Equipment

Crower introduced its Pontiac-spec Big Block Sportsman rod around 2000. Produced in America and constructed of 4340-steel, its unique I-beam design offers excellent strength and durability. It's popular with serious performance enthusiasts for engines producing up to 1,000 hp and turning as much as 8,000 rpm. A complete set sells for less than $800.

CP-Carrillo

CP-Carrillo's Pro-H series rod is among the most durable American-made options available today. Its standard H-beam, which is constructed of 4330-spec alloy is suitable for engines producing 1,000 hp and turning as much as 8,500 rpm, and fastener upgrade increases that to 1,300 hp and to as much as 9,500 rpm. CP-Carrillo can produce custom rods for engines producing 2,000 hp or more. Expect to spend around $2,300 for a complete set.

Oliver Racing Parts

Oliver is another premium connecting rod manufacturer. Its American-made rods are CNC-machined from a billet of 4340-steel. The unique I-beam offers a great deal of strength for its weight. Oliver rates a typical Pontiac-application rod up to 2,000 hp or slightly more, and it can accommodate any custom specification. Expect to spend at least $1,300 for a complete set of Oliver rods, and as much as $1,800 for a set of custom units.

Aluminum Rods

Since aluminum weighs less than steel, an aluminum rod can be made much beefier than a similar steel rod without the negative effects of excessive weight. This greatly improves durability while the lightweight construction reduces rotating mass, which can allow an engine to accelerate quicker, most likely improving performance.

The soft aluminum absorbs compressive shock loads. While that can lengthen cast crankshaft and/or bearing life in extreme applications, it, along with heating and cooling cycles, causes the alumi-

When it comes to steel connecting rods for a max-performance Pontiac V-8, Oliver's billet I-beam may be the best. Available in a wide variety of length and crank and wrist pin configurations, Oliver's rods are the choice of many professional Pontiac builders for high-end applications.

num to work-harden over time, making it very brittle and potentially shatter. Replacement is eventually required, and that could be 50 drag strip passes or 500, depending upon the particular application. That makes aluminum rods very impractical for a street engine, and though some have successfully done so, I don't recommend it. They are best suited for race engines that are torn down for freshening at regular intervals.

CHAPTER 4

Pressed Versus Floating Wrist Pins

An engine's piston is secured to the connecting rod by a length of high-strength tubular steel called a "wrist pin." Most production Pontiacs used a "pressed" wrist pin design while a "floating" wrist pin is much more common with aftermarket components.

A pressed wrist pin is secured to the rod body by an interference fit. The pin was usually driven into place using a hydraulic press, but that process can gall or distort the rod if it's not performed correctly. Carefully heating the rod's wrist pin bore to a specific temperature is a much better option and far more common today. The heat expands the metal, allowing the wrist pin to easily slide into position. The metal surrounding the pin contracts, locking it into place, as the rod cools naturally. The piston then pivots freely on the pin, providing quiet and consistent operation.

A floating-pin design requires a "bushed" rod where the wrist pin bore of a pressed-pin rod is enlarged and a bronze bushing is installed. The wrist pin is secured to the connecting rod by retaining clips on either end of the piston. This design allows the piston and pin to "float" on the connecting rod, dissipating the total compressive load. It also prevents any chance of tempering the rod and compromising its integrity by improperly heating the wrist pin bore for a pressed pin, and the bushing is replaceable during future rebuilds.

The best wrist pin retention method for your particular application largely depends upon the pistons you choose. Some stock-replacement pistons lack the retaining ring groove, which requires the use of pressed pins. Aftermarket high-performance pistons almost always are machined with a ring groove.

Connecting rods are specific to the type of wrist pins being used. Pressed-fit pins rely upon an interference fit to remain stationary, and the connecting rod's wrist pin bore must be honed to a spec that's only a couple thousandths larger than the wrist pin (right). A bushed rod is used in conjunction with floating wrist pins. A bronze bushing is pressed in place in such examples (left).

A length of tubular tool steel, known as a wrist pin, secures the piston to the connecting rod. It can be pressed in place or retained using locking clips on both ends. The wrist pin retention method you choose requires matching pistons and connecting rods.

GRP Connecting Rods

GRP produces a premium billet-aluminum I-beam rod that's suitable for a max-performance Pontiac V-8, particularly one that operates at very high RPM or uses forced induction. Weighing about 200 grams less than a similar steel unit, the GRP custom connecting rod for a Pontiac engine turns up to 8,500 rpm and produces as much as 1,200 naturally aspirated horsepower or up to 1,800 with a power adder. GRP can produce rods for engines with much more horsepower on a custom-order basis. Expect to spend around $1,200 per set.

For the alloy used in its rods, GRP typically recommends allowing around .010 inch more deck clearance when

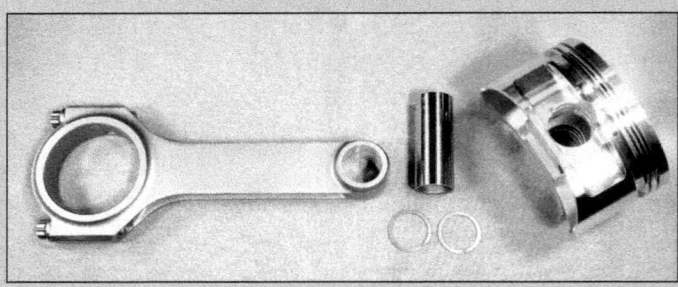

GRP produces some of the best billet-aluminum-alloy connecting rods available today. In addition to weighing substantially less than steel, aluminum also acts as a shock absorber between the piston and the crankshaft. This is particularly beneficial in applications where very high cylinder pressure exists. Aluminum fatigues and cracks over time, however, and rod failure is quite possible if the entire set isn't periodically inspected or replaced.

CONNECTING RODS

To positively locate the connecting rod body and cap, and to prevent the components from "walking," premium aluminum connecting rods feature radial serrations, such as those found on the GRP aluminum units.

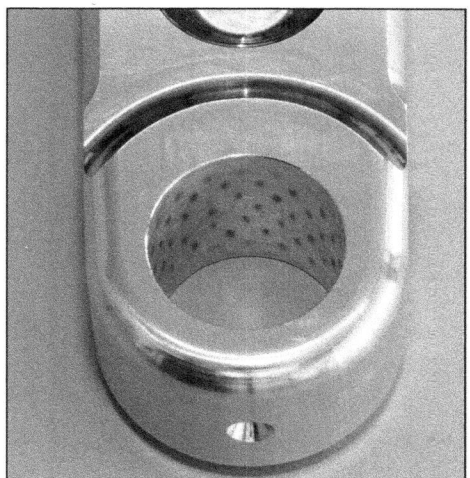

Aluminum connecting rods are not compatible with press-fit wrist pins. The aluminum alloy such rods are typically constructed of generally makes an excellent bearing surface. Additionally, the thermal expansion associated with aluminum makes it difficult to retain a bronze bushing.

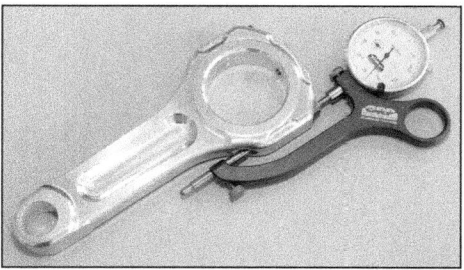

Any fastener applies maximum clamping load when it's stretched to a particular length. Stretch is often estimated by how much torque is applied to it. While that's suitable for fasteners that thread into a blind hole, it's easy to measure connecting rod bolt stretch during installation to ensure proper clamping load. Many companies offer bolt-stretch gauges, but this premium unit from GRP is compatible with any type of connecting rod.

compared to steel forgings, and an additional .001 inch of side clearance. That amount may vary with other aluminum rod manufacturers.

Connecting Rod Length

The stock Pontiac block's relatively tall deck height allows the use of a rather long connecting rod with a center-to-center length of 6.625 inches, producing a good rod-to-stroke ratio. A long-length rod tends to lessen the amount of side load placed on the cylinder wall, reducing cylinder wear and operational friction associated with a 90-degree V-8.

A common modification in modern performance rebuilds is to use a rod measuring 6.7 to 6.8 inches long. That length is specifically chosen for a few distinct reasons. It improves the connecting rod to crankshaft stroke ratio beyond stock, further reducing the significant cylinder wall loading that can occur when using an aftermarket crankshaft with a stroke length that's greater than 4.21 inches. It is also a length common to big-block Chevy applications, and easily fits into a Pontiac block. It does, however, typically require a corresponding piston, and crankshaft rod journals undersized to 2.2 inches.

By itself, I don't believe there's a measureable performance difference by simply increasing rod length from the stock dimension of 6.625 to as much as 6.8 inches. It does, however, allow the use of shorter, lighter-weight pistons, and does slightly decrease piston speed, and that's where its performance benefits may come from.

Fasteners

A connecting rod fastener is designed to provide a specific amount of clamping load and secures the rod cap to the body. That force is applied by stretching the bolt a very specific amount. Fastener manufacturers can accurately predict the amount of bolt stretch by how much torque is applied to it. While that's an acceptable method for most other fasteners, connecting rod bolt stretch can be precisely gauged using a specially designed fixture that measures a bolt's length from each end.

New rod bolts are commonly torqued to spec without measuring and/or affecting reliability, but measuring stretch is preferred for competition engines. During teardown, I highly recommend physically measuring a used rod bolt's length fully torqued and in a relaxed state. By comparing your recorded measurements to the fastener manufacturer's stated tolerances, you can assess a bolt's condition to determine if it's reusable. Keep in mind that a variance as small as .0005 inch can be the difference between reusing and replacing, so accuracy is extremely important.

There are two distinct types of connecting rod bolts available today. A thru-bolt type is a two-piece design that includes a nut and bolt. The bolt is pressed into the rod body and the nut secures the cap from the opposite end. A cap-screw type is a one-piece bolt that passes through the connecting rod cap and threads directly into the body. Thru-bolts are common in OE rods and

A two-piece connecting rod bolt is very strong, but requires a pair of opposing flanges that create the machined ledges for the nut and bolt to clamp against. The rod is usually thinnest at these points, and that creates weak areas where the rod can fail if pressed to its limits. Pontiac addressed this when designing its cast rod. Additional material around the flanges improves overall strength.

To maximize durability, nearly every aftermarket connecting rod manufacturer uses ARP fasteners in its rods. One-piece rod bolts are most popular since they occupy less space and leave more material throughout the rod. ARP offers a wide variety of one-piece replacements for use during subsequent rebuilds. Be sure to contact the connecting rod supplier for the correct ARP part number for use with its rods.

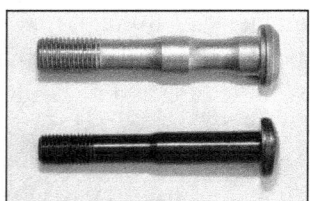

Original Pontiac connecting rods use two-piece rod bolts. ARP offers a few different high-performance replacements. Its 7/16-inch replacement is ideal when reusing original SD-455 rods. ARP also offers a 3/8-inch replacement for the original fasteners used with Pontiac's cast rod. Plan to have your machinist perform the installation, which usually includes light machining or honing.

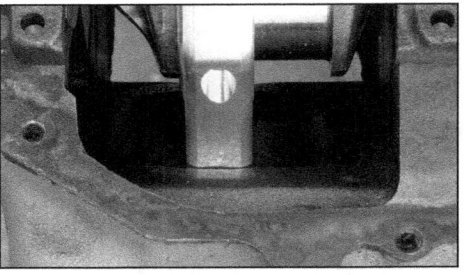

Because aluminum is not as strong as steel, aluminum connecting rods must be bulkier to provide a durability level similar to forged units. Depending upon the amount of crankshaft stroke, some areas of the block (particularly the bottom of the cylinder bore) may need to be relieved with a grinder to gain sufficient rotational clearance.

Before placing any connecting rod in service, have your local machinist verify that the crank and wrist pin bores measure within the specified tolerances. In most instances you should find nothing major. Your machinist can easily hone them to the required specification if any small variance is found.

As with each component of the rotating assembly, the connecting rods should be balance matched to ensure they all weigh the same. While some machinists simply weigh the entire rod on a scale, better shops have bob equipment that measures both ends of the connecting rod for maximum accuracy.

entry-level forged replacements. Cap screws are much more common in aftermarket offerings because they require less space within the rod, maximizing strength, and can be replaced easily.

Automotive Racing Products (ARP) is a leader in the fastener industry. It produces a number of top-quality connecting rod bolt kits, and most aftermarket connecting rods are already fitted with ARP bolts. I prefer ARP connecting rod bolts whenever replacement is required.

Selection, Preparation and Inspection

No matter the type, any rod that operates close to its yield strength can fail much sooner than one that's better designed to handle greater loads. Beyond improper installation or lubrication issues, most failures of an otherwise normal connecting rod are typically the result of an improperly chosen rod for the intended application. You can't always blame the rod manufacturer or installer if the connecting rod you selected isn't capable of handling the load associated with your particular engine. With that in mind, there's generally no harm in using a rod that's too strong for a given application, especially if you plan to increase its performance down the road.

Once you purchase the appropriate set of rods for your particular rebuild, thoroughly clean them and have your machinist verify the critical measurements of the entire set before placing them into service.

The critical dimensions should be rechecked each time an engine is torn down for freshening. That includes physically measuring for deformation, bending, twisting, or stretching, and even magnetically inspecting it for cracks. The rod bolts should be measured for stretch and replaced if necessary, and the bearings should be inspected for signs of abnormal wear.

CHAPTER 5

PISTONS AND RINGS

Auto manufacturers used cast-aluminum pistons in production engines for many years. They are cheap to produce, and they operate quietly and consistently throughout an engine's lifetime when properly designed. The piston that Pontiac developed during the late 1960s was an excellent piece. TRW copied its design to produce high-performance aftermarket forgings.

Pontiac specified forged-aluminum pistons for its Super Duty and Ram Air IV engines. The SD-455 piston was a unique design specific to that application. It contains many features aimed at maximizing durability; among them is a single-trough valve pocket, which displaces just over 3 cc. Federal-Mogul offered this piston (number L2423F) in various oversizes in its Speed Pro line, but was discontinued in February 2007.

Most Pontiac V-8s were originally fitted with cast-aluminum pistons, and they operated quietly and consistently. Forged-aluminum pistons were specified to improve the reliability and durability of Pontiac's very high performance engines such as the Ram Air IV and Super Duty 421 and 455.

Pistons

The cast Pontiac piston is a durable design that offers reliable performance throughout the lifespan of a typical engine. Its wrist pin is biased toward the thrust side to lessen cylinder wall loading and promote quiet operation. Large valve pockets allow maximum piston-to-valve clearance. A dish was machined into the crown of some to reduce compression ratio in certain applications. Though suitable for everyday driving, cast pistons are susceptible to failure when exposed to very lean conditions, extreme heat, and/or detonation over long periods of time.

Pontiac sourced forged-aluminum pistons from Mickey Thompson for the early Super Duty engines while TRW provided the forgings for the R/A IV and SD-455. Forged-aluminum pistons are far more forgiving than cast units and the

CHAPTER 5

When compared to the modern Speed Pro piston (right), it's easy to see how much influence Pontiac's original cast piston (left) had on the design of the TRW original stock replacement forging. Federal-Mogul purchased TRW in 1992 and began producing the pistons in June 2001, incorporating some modern technology in the process. The most apparent changes are in the skirt area.

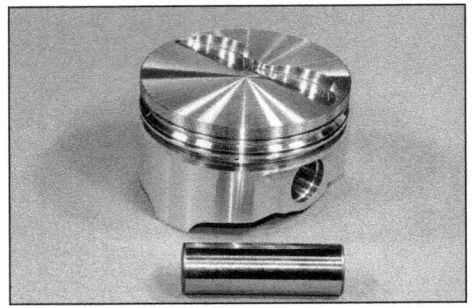

Keith Black (KB) offers a number of off-the-shelf forged pistons for Pontiac V-8s in its ICON line. Readily available from mail-order companies or Pontiac parts suppliers for about $400 and constructed of low-silicon 2618-alloy with thin ring packages, KB pistons are an excellent choice for any serious performance application. They are available with press-fit or floating wrist pins for stock-length or longer connecting rods, and with a variety of valve pocket volumes.

For those using an aftermarket block with a bore diameter of 4.35 inches, KB offers at least one off-the-shelf forged piston for 505-plus-ci engines in standard size, and in .005- and .010-inch oversizes. KB number-IC631 is designed for use with 6.8-inch connecting rods and features a compression height of 1.29 inches. The valve pocket measures just over 15 cc, which achieves a compression ratio of 10:1 when using 87-cc cylinder heads on a 505-ci Pontiac V-8.

TRW pieces are quite durable. Using the R/A IV design as the basis, TRW began offering a wide array of stock-replacement forgings for most Pontiac engines in stock displacement and popular overbore sizes. Several of the high-volume offerings remain available under Federal-Mogul's Speed Pro line.

Myriad stock-replacement cast pistons were produced throughout the years, and a variety is still available on the aftermarket at a very reasonable cost. A hypereutectic piston is made from a modern silicon-aluminum alloy that's often used to cast pistons, and it's generally considered a step better than aluminum. While pistons constructed of either material can certainly provide a long service life in a naturally aspirated engine, operating conditions must be ideal at all times or failure can result. With that, I am not comfortable using cast or hypereutectic pistons in any high-performance engine.

I consider forged-aluminum pistons a must in any performance engine, especially if nitrous oxide or forced induction is planned or could be used in the future. Forgings are generally much more tolerant to high-combustion heat, momentary lean conditions, and/or detonation. While no piston should be expected to survive indefinitely if such conditions aren't quickly corrected, there's much less chance of failure with a quality forging when compared to a cast piston.

Off-The-Shelf Pistons

Most Pontiac vendors can supply you with a set of high-quality forged pistons at a price ranging between $400 and $600. JE Pistons, Keith Black, Probe, and Speed Pro are just a few companies that produce off-the-shelf forging that fit the Pontiac V-8 without modifications. Options can range from a stock-replacement design to a lightweight forging better suited for high-RPM applications.

JE Pistons offers several high-quality flat-top pistons for 400, 428, and 455 engines in its SRP line. Constructed of high-silicon aluminum alloy, its pistons are fully machined to a limited number of specific overbore sizes and feature a valve pocket displacing 5 cc. The wrist pin hole is

Machining a dish into the piston head is an accepted way to reduce an engine's compression ratio. The piston being machined must have enough material in its head to accommodate that dish while maintaining optimal strength for the application, however. KB offers a limited number of forged pistons featuring a machined dish that measures between 10 and 16 cc, depending upon the application. The dish of this particular example measures 15 cc. It is an excellent choice when building an engine intended to operate on pump gas that uses small-chamber cylinder heads.

bored to accept a .980-inch-diameter pin for use with stock-length Pontiac connecting rods for certain applications or .990-inch-diameter pin for stroker engines using

PISTONS AND RINGS

When Federal-Mogul purchased TRW, it incorporated TRW's Powerforged line of high-quality forged pistons into its own Speed Pro catalog. Though many low-volume Pontiac offerings have been discontinued, a number of stock replacement forgings, such as this +.030-inch 455 example (number L2359F030), are readily available in many popular original Pontiac sizes, and in oversizes common for rebuilds. Expect to spend about $400 for a complete set.

6.8-inch rods. The wrist pin hole is grooved to accept lock rings for floating pins.

The ICON line by Keith Black contains a number of high-quality forged pistons for original 400 to 455 Pontiac blocks, and at least one large-bore aftermarket Pontiac block dimension. Available in myriad standard and overbore sizes, and in a flat-top design with a valve pocket displacement of 4 to 5 cc, or a dished crown ranging from 10 to 30 cc to reduce compression, ICON pistons are constructed of a low-silicon alloy aluminum for maximum endurance and reduced drag. The wrist pin hole is machined for floating pins and accepts .980- or .990-inch pins depending upon the application.

Federal-Mogul offers several high-quality forged pistons in its Speed Pro Powerforged line. Constructed of extruded aluminum alloy, an updated skirt design reduces drag and is moly coated to resist scuffing and reduce side clearance for quiet operation while the 5/64-inch top ring provides maximum

Calculating Piston Compression Height

A piston's "compression height" is the distance from the center of the wrist pin bore to the deck surface of the crown. It usually measures between 1.48 and 1.70 inches on a typical Pontiac engine that uses stock or stock-replacement components. An engine built with longer crankshaft strokes and/or connecting rods may require a shorter compression height, such as 1.3 inches.

Positioning the wrist pin higher in relation to the piston's deck surface, which reduces compression height, allows for the use of a longer connecting rod, which improves rod-to-stoke ratio. The required compression height for any piston can be calculated by using the following mathematical formula that factors in several common engine measurements:

Compression Height = deck height − half of crank stroke − rod length − deck clearance (or how far the piston is recessed in the bore)

The compression height can be reduced to 1.2 inches or less, but that can present some issues that must be considered. The wrist pin bore is so high that the oil control ring groove opens into it. That requires a special oil-control ring package that includes a support rail that must be installed below the oil ring. Severe piston rock can also occur when the compression height is less than 1.2 inches. That can lead to scuffed piston skirts and/or cylinder walls. If long-term reliability is important, a compression height of at least 1.3 inches may best provide it.

Compression height is the measured distance between the centerline of the wrist pin bore and the deck surface of the piston head. Compression height varies greatly from engine to engine. Block deck height, crankshaft stroke, and connecting rod length are the factors that determine the required amount. This particular KB piston has a compression height of 1.49 inches.

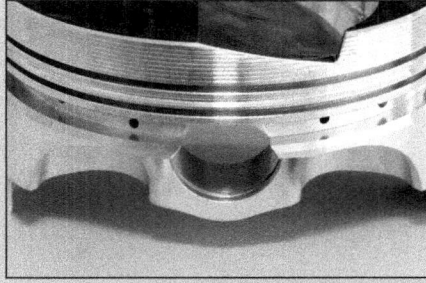

When compression height nears 1.2 inches, the wrist pin bore is so high that it oftentimes breaks into the lower rail of the oil ring groove. That requires a special piston ring set that includes an oil ring support rail, which is a thick "shim" that replicates the missing portion of the rail, effectively supporting the oil ring during operation. Lack of oil ring support can lead to excessive oil consumption and various other abnormal operating conditions.

reliability and cylinder seal. Considered a high-performance stock replacement for select 389 to 455 engines in limited stock and overbore sizes, the crown area is very thick for maximum durability, and can also be machined for custom applications. Because the valve pocket is positioned for valves in the stock location, some modification may be required when using certain aftermarket cylinder heads and/or larger-diameter valves. The wrist pin bore is offset as on a Pontiac original, and is machined for a press-fit .980-inch wrist pin only.

Custom Pistons

If your rebuild uses a stock-stroke crankshaft and stock-length connecting rods, then stock-replacement forgings are an excellent choice. Custom pistons are sometimes required when using a long-stroke crankshaft, aftermarket connecting rods, specific overbore sizing, or for various other reasons. Diamond, Ross, and Venolia are a few companies that produce high-quality custom forgings. When designing a custom piston, the wrist pin bore, ring grooves, and valve pockets can be located in most any position on the piston.

Some Pontiac builders regularly stock specially designed custom pistons for their proprietary stroker kits. Special characteristics may be required if custom pistons are required for your particular Pontiac. Expect to spend more than $600 for a quality set. If you plan to tear your engine down regularly, you might consider ordering one or more extra pistons. Having spares can save you time and money.

One custom piston feature that's somewhat as popular with other makes, but not always popular for Pontiacs, is gas ports. Gas ports are a series of small holes that are precisely drilled into the head of the piston and through to the ring groove. A gas port allows a small amount of cylinder pressure to pass behind the top ring, pressurizing it and improving cylinder seal. It can be beneficial when using very narrow ring packs and/or high ring positions, which are commonly associated with very high RPM engines. If you plan for such operation with your Pontiac and will be using custom pistons, it may be a feature to consider.

Piston Rings

Piston rings are intended to keep cylinder pressure from entering the crankcase while maintaining a very light film of oil on the cylinder walls to lubricate the piston as it travels up and down. Most Pontiac pistons utilize a ring pack consisting of three separate piston rings. The top two are compression rings, which serve to keep cylinder pressure from entering the crankcase while scraping excess oil off the cylinder walls to reduce oil consumption. The bottom ring is designed to keep a majority of the engine oil that's splashing about the crankcase (and onto the cylinder walls) below the compression rings, thoroughly lubricating the piston skirt in the process.

Generally speaking, improved piston ring seal equates to greater horsepower and torque. Modern technology has improved piston ring design. Available materials and cylinder-wall finishing processes have improved the performance and longevity of an otherwise typical piston ring set. Compression rings can be

When the required dimensions of a piston for a particular engine vary beyond what's commonly available in off-the-shelf sizes, companies such as Ross Racing Pistons specialize in producing high-quality forgings that feature custom dimensions. Ross can accommodate most requests, offering many unique options to provide you with the best piston possible.

Total Seal is an industry leader in piston ring design and versatility. From barrel-faced, ductile iron conventional rings with a moly-faced coating to high-end diamond lapped rings with gapless ends, it offers many common off-the-shelf sizes and produces myriad custom sets as well. Total Seal offers some of the best ring sets in the industry, especially for highly specialized applications where very high cylinder pressure or very low drag is required.

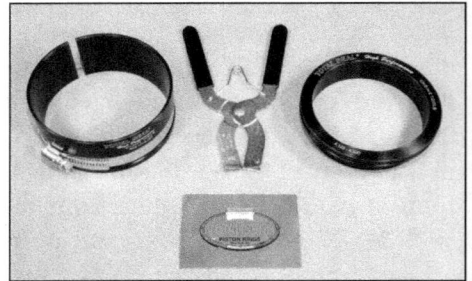

Piston rings are rather delicate and proper installation is critical to operational consistency. Total Seal offers a number of tools that allow correct and easily installation. That includes a specific tool for compressing rings when installing assembled pistons into the block (left), installing rings on pistons (center), and a squaring tool (right) for test fitting rings in the block to measure end gap. Such tools are readily available from most mail-order suppliers and premium local auto parts stores.

PISTONS AND RINGS

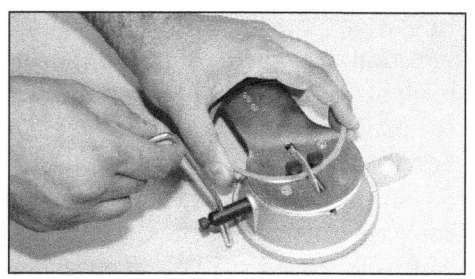

A specific amount of piston ring end gap is required to prevent the ring ends from butting against one another as the ring expands as the engine comes up to temperature. Most piston rings must be trimmed to achieve the recommended amount of end gap for a particular application. A small file locked in a vise is acceptable. The ring ends should be filed carefully and slowly in a pattern from the face toward the center to prevent chipping any face coatings. I prefer a hand-operated end gap grinder such as this from Federal-Mogul (number MT-135). It grinds easily and consistently and has a replaceable wheel. No matter what method you use to adjust the end gap, be sure the ring ends are parallel to prevent any operational issue.

Piston Ring End Gap Guidelines

Federal-Mogul offers a simple equation to determine the end gap requirements when combining its Speed Pro line of high-performance piston rings in a variety of engine types that operate at normal temperatures. You likely notice that the end gap recommendations increases along with performance potential. High-performance engines typically produce greater horsepower and heat, and the wider piston ring end gap recommendations compensate for higher rates of thermal expansion.

Consider the measurements below as a guideline to help you determine the correct end gap range for your particular application.

Speed-Pro Recommendations*

For Moderate Performance
Top Ring   .004 inch
Second Ring   .005 inch

For Drag Racing
Top Ring   .0045 inch
Second Ring   .0055 inch

For Nitrous Oxide, Street
Top Ring   .005 inch
Second Ring   .006 inch

For Nitrous Oxide, Race
Top Ring   .007 inch
Second Ring   .007 inch

Turbo or Supercharged
Top Ring   .006 inch
Second Ring   .006 inch

* per inch of bore diameter

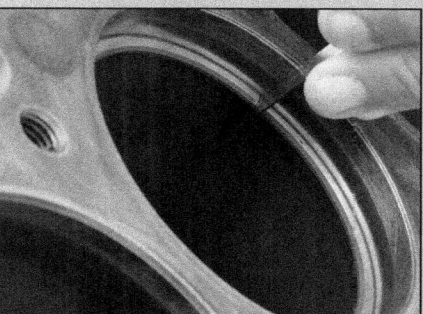

Measure the end gap of each compression piston ring in its particular cylinder. Ideally, a torque plate should be used, but that's not always possible or practical. Lightly coat the ring and wall with engine oil to prevent scratching or scuffing, and use a piston ring squaring tool to be sure the ring is correctly located for measuring. A feeler gauge is then used to measure the amount of end gap and the ring is trimmed accordingly for the application.

made much thinner, significantly reducing operational friction without compromising cylinder seal, and applying this technology to your Pontiac has benefit.

When selecting a piston ring package, you will find that most aftermarket forged pistons require 1/16-inch-thick compression rings, and a 3/16-inch-thick oil-control ring. Most conventional ring sets are purposely left a bit tight and must be file-fit by hand to achieve the proper amount of end gap for a particular combination. Some are even designed to run gapless. Be sure the ring type that you select is compatible with the cylinder wall finish left after honing.

I have used various Speed Pro ductile iron rings with moly-face coatings with excellent results. Total Seal offers a variety of readymade piston ring kits for many popular off-the-shelf Pontiac pistons, and it can quickly produce custom rings in any width or thickness for highly specialized applications, such as those running nitrous or forced induction. I highly recommend discussing your piston ring options with the piston manufacturer and/or your Pontiac builder to ensure you're making the best selection for your particular application.

Piston Ring End Gap

A specific amount of compression ring end gap is required during assembly. It allows the ring to expand controllably in the cylinder as the engine reaches normal operating temperature. Too much end gap can allow an excessive amount of cylinder pressure to pass into the crankcase—a performance-compromising condition referred to as "blow-by." An end gap that's too tight can cause the ring to bind with the cylinder bore, scoring the cylinder wall, and breaking the piston ring and/or ring lands.

In years past, it wasn't uncommon to find end gap specifications wider for the top ring than the second. Generally speaking, modern ring packs are designed to operate with about .004 inch of top compression ring end gap for each inch of bore diameter, and .005 inch of second compression ring end gap for that same measurement. The oil-control ring gap simply needs to be more than .015 inch in nearly all applications.

Preparation and Inspection

There are many excellent piston options. Once you've made your selection, it's wise to closely inspect each new piston before sending the entire set to your machinist for proper weight balancing along with the rest of the rotating assembly. For me, that process typically includes taking very fine sandpaper (1,000 grit or more) that's been wetted to the valve pocket area in the head of the piston. A red scouring pad also suffices.

The idea is to gently remove any sharp edges that could otherwise induce detonation if the fuel octane is ever marginal for the engine's compression ratio, or if its operating temperature is ever unusually high. I run my finger over the ring lands and carefully use the same wetted sandpaper to remove any burrs that might scratch the cylinder wall finish and/or prevent the piston rings from properly sealing. I also recommend inspecting the supplied wrist pin to be sure it fits tightly and can easily be pushed from its bore

Assuming that sufficient cylinder wall clearance was present at all times and there were no lubrication issues or major failures, you will likely find the pistons in excellent condition during routine teardowns. If you hope to reuse the pistons, the first step should always include measuring the dimensions with a micrometer. If any of them is not within the manufacturer's stated tolerances, it's best to replace the entire set.

If your pistons and block measure correctly, then there shouldn't be any issue with reusing the pistons once properly cleaned. Any carbon built up on the head can be removed chemically. It's best to ask your piston manufacturer for its recommendations on what to use. It isn't unusual to find some very light scratches on the piston skirts during the inspection process, particularly on the thrust side.

You might be tempted to reuse pistons that are slightly loose, but in otherwise good condition. This, however, can lead to excessive piston skirt noise (piston slap), wrist pin noise, and/or oil consumption. In the past, knurling was a popular way to reuse slightly-worn pistons. The knurling process gathers material around the skirt area and expands the piston's dimension to compensate for the wear. I

When pistons are machined, the process may leave some sharp edges behind. Though rare, any sharp edges created when cutting the valve pockets can create hotspots that induce detonation in certain operating conditions where fuel octane is marginal. To prevent any chance of detonation related to it, I routinely remove any sharp edges prior to installation with 1,000-grit sandpaper that's been wetted. Little more than a gentle polish is all that's required.

Before placing a new piston set in service, plan to have your machinist verify the diameter of the wrist pin bore of the entire set. The intent is to ensure that it's within the suggested tolerance with the wrist pins being used. Don't be alarmed if some minor honing is required, especially if floating pins are used.

PISTONS AND RINGS

Since the pistons are part of the rotating assembly, they too must be balance matched along with all other components for optimal engine reliability and performance. Most manufacturers leave a "pad" of material located just below the wrist pin bore. Material is removed from this area until all the pistons of a particular set weigh the same.

Lock rings are used to retain floating wrist pins, and two are generally used on each end. The rings are coiled much like a key ring, and proper installation consists of gently stretching the lock ring to separate the coils and then placing the lock ring in the corresponding groove that is machined into the piston and gradually walking it around until it snaps into place. New lock rings should be used during each teardown.

Hobbyists have found several ways to install piston rings. Some cause permanent distortion or a loss of tension, however, which greatly affects its sealing ability. After end gap has been measured and/or adjusted accordingly, a compression ring is properly installed onto a piston by spreading it with a special tool until it just fits over the head and then is carefully slipped into place. Remember a top ring is often gapped differently than a second, and some ring kits require special orientation, so plan accordingly.

feel this practice has no place in a modern performance engine.

Piston Coatings

Engine component coatings are popular in today's performance world, and pistons are likely the most commonly coated engine component. Some coatings, such as a ceramic applied to the head of the piston, act as a thermal barrier that rejects heat. It can increase the amount of combustion heat a particular piston can withstand, while preventing the heat of the combustion process from escaping, potentially improving power output. Other coatings, such as a dry film moly-based lubricant applied to the skirt, serve to reduce friction and wall clearance while improving acceleration without skirt damage.

Certain Pontiac pistons, such as the Speed Pro units from Federal-Mogul, come with a light moly-based coating that's factory applied to the skirts. It reduces friction and provides quieter operation when compared to non-coated examples.

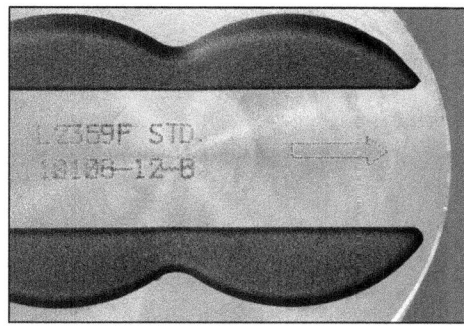

If a piston's wrist pin bore is offset, there should be an arrow or notch indicating which end should face forward during installation. The valve pocket of some pistons is shaped in a way that the intake valve pocket is larger and/or deeper than the exhaust, and the pistons must be oriented accordingly. Pay particular attention to head design of the pistons used during your build to be sure they are installed correctly. Consult the piston manufacturer if you have any questions.

Many independent tests indicate that engine coatings improve performance in certain applications. Ceramic coatings applied to piston heads reject heat, while dry film coatings applied to the piston skirts reduce rotational friction and improve component longevity. Federal-Mogul coats the skirt of many pistons in its Speed Pro line with moly-graphite, which acts as a wear lubricant to protect scuffing on initial startup and ensure quiet operation. Many companies specialize in such coatings, and I suggest asking your engine builder to determine if any coating is recommended for your particular build.

CHAPTER 6

CYLINDER HEADS

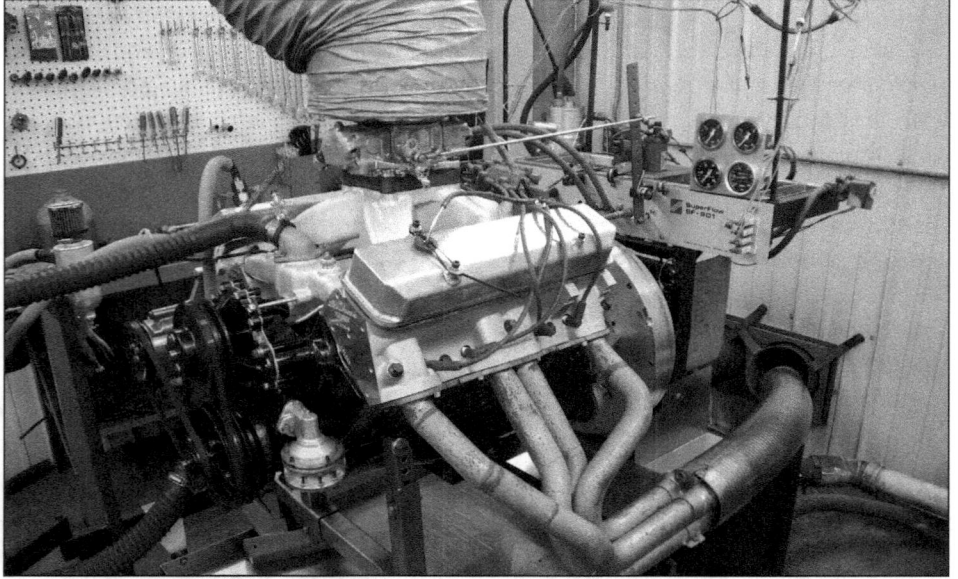

An internal combustion engine is basically an air pump. Improving its ability to move a greater volume of air almost always increases output in a particular range. Selecting a cylinder head with the proper amount of intake and exhaust airflow and volume generates the greatest amount of horsepower and torque over the intended operating range.

In its simplest terms, an internal combustion engine is nothing more than an air pump. The more air that's drawn into a cylinder as the intake valve opens and closes, the more intense the combustion event is when the appropriate amount of fuel is added, and that translates into greater potential for horsepower and torque. A given engine can only use so much airflow. Just how much depends upon displacement, but maximum operating speed and operational efficiently are most important.

Intake Airflow

Peak intake airflow is a common reference when comparing cylinder heads. While that figure is very important, especially for attaining strong peak horsepower, a cylinder head spends very little time at peak valve lift and considerably more time reaching that point. Considering airflow throughout the entire valve lift range is more important. Strong low- and mid-lift airflow starts cylinder fill quickly and effectively and continues filling the cylinder until the intake valve closes.

SuperFlow Corporation is an airflow industry leader and it has developed a mathematical equation that can predict approximate horsepower based upon airflow at a specific test pressure. When discussing peak airflow of a cylinder head today, it's usually stated at 28 inches of water pressure. Though SuperFlow lists formulas for all common airflow test pressures, I am including only the one for 28 inches since it's a pressure most are familiar with.

Horsepower per Cylinder = (Cpower x Peak Airflow) x Number of Cylinders

The equation arrives at peak horsepower per cylinder by multiplying peak airflow by the SuperFlow's predetermined "Coefficient of Power" (Cpower). At 28 inches of pressure, Cpower is equal to

CYLINDER HEADS

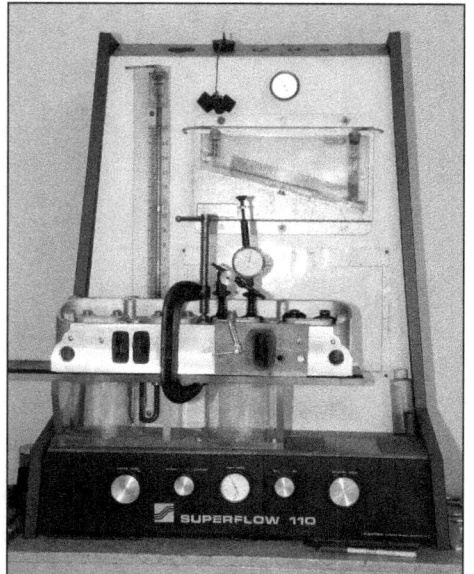

A flow bench is used to measure the intake and exhaust airflow of a cylinder head. A typical cast-iron Pontiac D-port measures 190 to 210 cfm of peak intake airflow at 28 inches of pressure. Most aftermarket aluminum castings flow anywhere between 260 and 330 cfm. Specialized castings can flow more than 400 cfm.

.260. You must then multiply the result by the number of cylinders for the subject engine, 8 in the instance of the Pontiac.

If your cylinder heads are capable of providing 240 cfm of peak intake airflow at 28 inches of pressure, the engine could produce as much as 500 hp.

Simply reversing the formula can determine approximately how much peak airflow is required to achieve a target horsepower.

Peak Intake Airflow = (Target HP ÷ Number of Cylinders) ÷ Cpower

If 600 hp is your goal, for instance, it requires an intake port capable of flowing nearly 290 cfm.

SuperFlow's equation is reasonably accurate. You may have noticed that it does not consider engine displacement. That's because peak horsepower is derived

It's easy to determine which of the ports are larger. A smaller port flows well for strong peak horsepower, but maintains good port velocity for quick throttle response and low end torque. A larger port provides much more peak airflow while delivering a greater volume of air to engines that operate at high RPM. The tradeoff is less port velocity at low speed.

from engine speed, and not size. A small-displacement engine can make an equal amount of horsepower as a larger engine using the same available airflow, only its horsepower peak occurs at a much higher RPM.

Port Volume

A greater volume of airflow is required as engine speed increases. The overall size of an intake port is referred to as its cross-sectional area. It's calculated by multiplying the recorded height and width at particular points within the port.

When comparing cylinder heads where intake port length is nearly identical, port volume is a common figure that lends a general insight to the cross-sectional area of a particular cylinder head. Generally stated in displaced volume of cubic centimeters, port volume is relative to engine size and can help determine how a particular cylinder head influences engine operation.

In general terms, a larger intake port flows a greater volume of air. While that can allow a given engine to operate effectively and efficiently at higher speed, the tradeoff is less port velocity. Port velocity is a critical element of engine performance. If you imagine airflow as a liquid, and relate it to water through a hose, the same volume of water exits a garden hose with far more velocity than it would a fire hose. Applying that concept to airflow passing through the intake ports of your Pontiac's cylinder head, it's easy to understand its effect on cylinder fill.

A smaller intake port is ideal for an engine that spends most of its time operating from idle speed to a moderate shift point such as a typical passenger car engine or a street/strip application where a broad power band is required for operation over a wide range. A larger port is ideal for a race engine that operates in a relatively narrow RPM range, and one where low-speed street manners are not a concern.

Smaller-displacement engines, particularly those with a shorter crankshaft stroke, are more sensitive to port volume. A larger engine such as the 455 simply moves more air per revolution than a smaller engine because of its size, and its longer crankshaft stroke tends to pull harder on the intake port. A larger engine can tolerate greater port volume without as many negative effects. That doesn't suggest that a port can never be too small or too large, however.

Exhaust Airflow

The byproducts of normal combustion must be efficiently expelled out the exhaust port during the exhaust stroke. Exhaust airflow is equally as important as intake airflow. A typical exhaust port isn't nearly as large as the intake port on a given cylinder head, and its valve isn't quite as large, so it doesn't have the capacity to move as much static airflow. That's not a significant issue, however.

Many builders agree that an exhaust port that flows 70 to 75 percent of the intake port throughout the entire lift

CHAPTER 6

Flow Benches

Cylinder head airflow is stated in cubic feet per minute (CFM) at a particular water depression. A flow bench is a professional tool commonly used to measure and record the intake and exhaust airflow capacity of a given cylinder head. Many different companies produce flow benches, but SuperFlow Corporation is an industry leader and produces entry-level, tabletop units for hobbyists and small shops, as well as much larger, standalone flow benches for professional engine builders and race shops.

Most professionals use the larger SuperFlow SF-600 or SF-1200 flow benches because they can flow up to a depression of 25 to 28 inches. Higher test pressures are no more accurate than lesser pressures. In fact, SuperFlow states in its operator manuals that virtually any test pressure can be used and mathematically converted to a higher test pressure with accurate results. The advantage of higher test pressure is that the operator can easier detect small airflow differences.

I have owned a SuperFlow SF-110 for several years and use it regularly to flow test various cylinder heads, intake manifolds, and carburetors. Depending upon the casting and valve lift, I use a test pressure between 10 and 15 inches. By carefully following SuperFlow's suggested bench preparation and testing procedure, I have repeatedly found that after mathematically converting my results recorded at a lesser pressure to a higher pressure, the numbers throughout the entire lift range are practically identical to those recorded at 28 inches on a larger bench.

A flow bench can be an excellent tool when comparing cylinder heads, but you must remember that the results can vary depending upon the bench and testing procedure. It is very easy to artificially increase or decrease the overall airflow capacity of a particular cylinder head by not calibrating the bench properly, not using the appropriate temperature differential, or simply being lazy when reading the pressure or flow manometers. If airflow numbers are important to you, I suggest purchasing cylinder heads from a reputable source with an accurate flow bench and proven performance results.

A cylinder is under pressure after the combustion event. As soon as the exhaust valve opens, exhaust gas begins venting out under its own pressure. The piston pushes the residual gas out. The exhaust valve and port do not need to be quite as large as the intake. An ideal exhaust port flows between 70 and 75 percent of the intake port throughout the entire lift range. A short extension is often used when measuring exhaust airflow on a flow bench. It replicates the first few inches of the header tube and can increase flow by several percent.

range is sufficient for maximum performance. For instance, an intake port that peaks at 300 cfm might perform best with an exhaust port that peaks at 210 to 225 cfm. Just how much exhaust flow is required largely depends upon the compression ratio of a particular engine.

The compressive force that occurs as the air/fuel mixture ignites pushes the piston downward, rotating the crankshaft, and ultimately motivating the vehicle. As the piston reaches bottom dead center (BDC) during the power stroke and begins traveling upward on the exhaust stroke, the exhaust valve opens and excess pressure trapped in the cylinder vents outward through the exhaust port. It's commonly referred to as "blow down."

Blow-down pressure is directly related to an engine's compression ratio. A lower ratio tends to produce less blow-down pressure, and because of this, a low-compression engine can usually benefit from greater exhaust airflow. An engine with a higher compression ratio has greater blow-down pressure, and air escapes more forcefully as the exhaust valve opens. It is this condition that gives high compression the crisp "pop" heard at the tailpipe.

Low- and mid-lift exhaust airflow is very important. The quicker blow-down pressure can evacuate the cylinder, the less negative work the engine must perform as the piston travels upward during the exhaust stroke. That doesn't suggest that peak isn't important. The exhaust port is only a small portion of the entire exhaust tract, and there are other factors to consider such as diameter of the header tubes and the exhaust system beyond it.

The power of a particular engine that is exhaust flow deficient tends to fall off rapidly. It's difficult for an exhaust port to flow too well, but it can happen. The velocity of the exhaust charge exiting the engine creates a vacuum that draws the exhaust from the adjacent cylinders and header tubes. If exhaust tubing is so large that the charge isn't able to maintain

CYLINDER HEADS

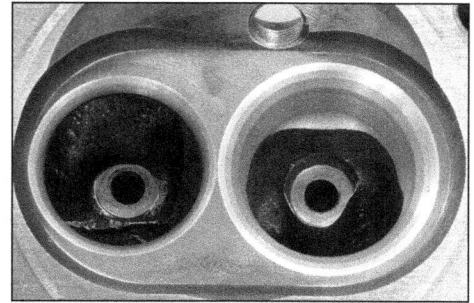

A high-performance valve job consists of multiple angle seat cuts that improve the airflow's transition from the bowl area into the chamber. Modern cutting bits can make several cuts at one time. A multiple-angle "valve job" should be part of any high-performance engine build.

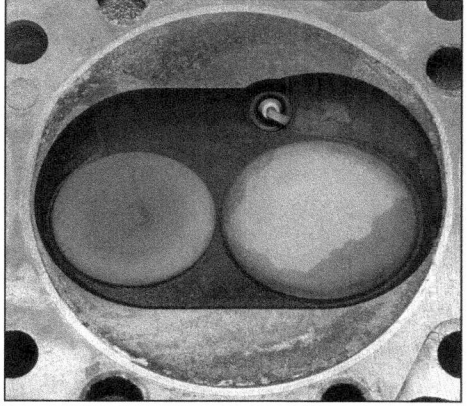

Through the 1967 model year, most every cast-iron Pontiac D-port featured a fully machined "closed" combustion chamber. This design was effective early on but was eventually replaced.

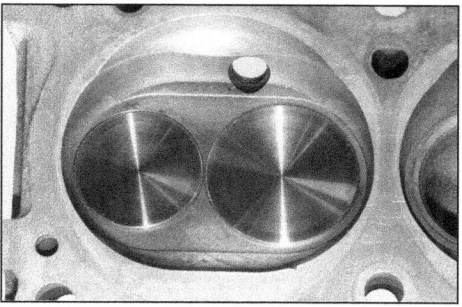

Pontiac began widely using a wedge chamber with an "open" design in 1968. It improved airflow dispersion and flame propagation. Whether it actually improved performance is debatable, but Pontiac found it was more efficient and easier on tailpipe emissions.

Modern production vehicles use a fast burn combustion chamber for its excellent efficiency. The spark plug is located in an optimal position for maximum burn. Most aftermarket companies have incorporated a "heart-shaped" chamber into the cast-aluminum Pontiac V-8 offerings available today. It typically requires less total spark lead to produce peak power.

its velocity, an engine with excessive exhaust flow revs lazily and lacks strong low-end and mid-range torque.

High-Performance Valve Job

A high-quality valve job that includes multiple seat cuts generally improves airflow at every lift point, and the use of specific angles can bolster airflow in specific ranges. Pontiac originally used a 30-degree intake seat angle on its regular production, large-valve cylinder heads as it tends to enhance low-lift airflow. A 45-degree intake valve seat angle was common in such high-performance applications as the Ram Air V and Super Duty 455, where maximizing peak airflow was important.

A valve seat angle can accentuate the way airflow transitions from the intake port into the chamber. Selecting the best seat angle for a particular cylinder head casting largely depends upon the shape of the overall port shape, valve bowl area, and combustion chamber shape. The smaller bowl area associated with cast-iron originals tends to favor a 30-degree seat angle while a wider bowl area, such as that of the Edelbrock RPM favors a 45-degree seat angle.

A seat angle greater than 45 degrees is quite common when building race engines of other makes. At least one professional Pontiac engine builder says the gains are limited unless valve lift exceeds .700 inch. The tradeoff for peak airflow is significantly reduced low- and mid-lift airflow. Unless the cylinder heads are extensively ported and the engine will operate in a relatively narrow RPM range, a valve seat angle in excess of 45 is rarely required.

Combustion Chamber

The combustion chamber is an extension of the valve seat and the seat angles. The multiple cuts associated with a high-performance valve job can smooth the transition and improve airflow. A properly designed chamber allows airflow to smoothly transition from the port into the cylinder. A combustion chamber that shrouds the valves can restrict airflow as it expands, limiting airflow at all lift points.

Original Pontiac cylinder heads contain a wedge-type combustion chamber that's fully machined for a clean, consistent burn. All modern production engines, and most aftermarket Pontiac cylinder heads, feature a fast-burn combustion chamber, which is easiest identified by its distinct heart-shaped appearance. Both are very effective at providing strong performance, but the fast-burn chamber is clearly more effective.

A fast-burn type of chamber is specifically designed to agitate the incoming air/fuel mixture, and the spark plug is strategically placed where combustion is most intense. The result is a highly efficient combustion event that, when compared to a typical wedge chamber, generally requires less spark lead to produce peak power, and can also require less fuel in the process. The latter is ideal for modern production engines where efficient combustion can reduce tailpipe emissions.

Compression Ratio

There are two different compression ratio ratings associated with engines: static and dynamic. A static ratio uses the physical measurements of a particular engine to mathematically calculate an actual ratio. It uses such information as bore and stroke, piston-valve relief, deck-height volumes, head-gasket thickness, and combustion-chamber volume. Any change to static ratio involves replacing one or more of the aforementioned components.

Dynamic compression ratio uses the same variables to calculate as static ratio, but it also factors in cylinder pressure. During normal engine operation, there is a brief overlap period where both valves are open as the exhaust stroke ends and the intake stroke begins, and the cylinder cannot begin to pressurize during the compression stroke until the intake valve closes. Both conditions have a significant effect on dynamic compression ratio. Advancing or retarding the camshaft's position in relation to the crankshaft will affect it.

Either compression ratio can be calculated using complex formulas. A quick search of the web provides several static ratio calculators on any number of auto enthusiast sites. Dynamic ratio calculators are more difficult to find, and require inputting detailed camshaft valve events. I commonly use the compression calculator on the Performance Trends Engine Analyzer program to calculate the static and dynamic ratios of various engines. It's amazing to watch how small changes can affect compression ratio.

When determining a target static compression ratio for an engine, there aren't any negative attributes associated with increasing compression ratio so long as sufficient fuel octane is available. Compression adds horsepower and torque throughout the entire RPM range and it can make a given engine seem larger at low speed and can make the camshaft seem larger at high speed.

A fuel octane rating of 91 is the highest commonly available at gas station pumps today. The greatest compression ratio I am comfortable suggesting when combining cast-iron Pontiac cylinder heads with 91 octane fuel is 9.5:1. Like many others, I have pushed that toward 10:1 or slightly more without any negative effects in most instances, but detonation can destroy a fresh engine in short order. The fuel and spark curves and cooling system must be constantly monitored and maintained. The extra effort and risk may not be worth the slight power increase that results.

Cast-aluminum cylinder heads are very popular in modern builds. Quicker heat dissipation is one major benefit that aluminum castings offer over iron. Modern GM service literature states that aluminum can dissipate heat 2.5 times quicker than gray iron. That allows a cast-aluminum cylinder head to tolerate more compression before detonation

Some of Pontiac's best flowing and most capable D-port cylinder heads were produced during the 1970s. They are low-compression castings, however, and have rather larger combustion chambers. Your machinist can mill the deck surface to reduce chamber volume. Because the Pontiac V-8 is a 90-degree design, the intake flange must be milled an equal amount for proper port alignment.

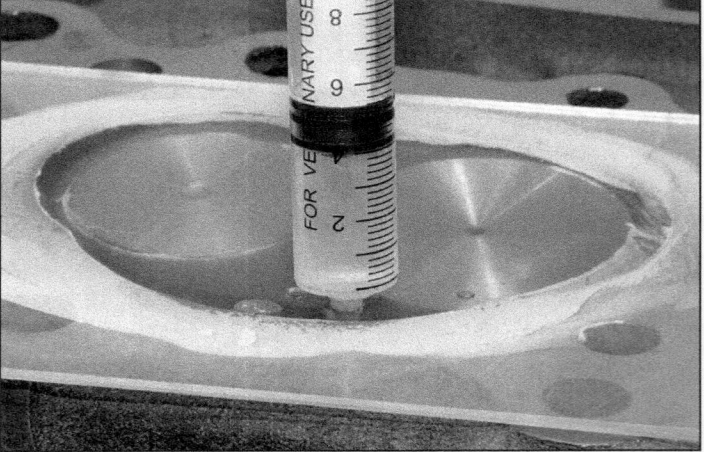

In general terms, each .005 inch removed from a typical open-chamber cylinder head reduces volume by 1 cc. Most agree that .060 inch is the most that can safely be removed from any unmodified casting. I have seen some milled as much as .1 inch for dedicated race engines, but I don't recommend that for everyone. Physically measure chamber volume before and after machining for maximum accuracy. Do this with a graduated cylinder, or even a veterinary syringe and rubbing alcohol.

occurs and most incorrectly assume that it automatically translates into a power increase. That's not totally true, however.

The thermal efficiency of aluminum is less than that of cast iron. An aluminum casting allows a slightly greater amount of heat from every combustion event to dissipate into the coolant jacket surrounding the combustion chamber, and less of it is applied as force on the piston. Because of this, an aluminum casting typically requires at least .500 to .750 more compression to make power that's equal to an iron casting. When combining a cast-aluminum Edelbrock RPM cylinder head with 91-octane pump fuel, I suggest a maximum target compression ratio of 10.25:1. The Performer RPM's wedge chamber isn't quite as efficient as a modern, fast-burn type found on other castings, and exceeding that ratio can induce detonation. If variables are closely monitored, I have seen some successfully push that toward 10.5:1, but once again, I feel the performance gain warrants the possible consequence.

Most every other cast-aluminum Pontiac cylinder head available on the aftermarket today features a fast-burn combustion chamber. These castings tolerate slightly more compression ratio. I recommend a target compression ratio of 10.5:1 when combining the Edelbrock Performer D-port or KRE D-port with 91-octane pump fuel. Some builders have reported successful results running a static ratio as high as 11:1, but I consider those specialized applications and certainly wouldn't recommend that for everyone.

A static compression ratio between 13:1 to 14:1 is quite common in race engines where 108- to 110-octane race fuel is used exclusively. It's possible to push compression toward 14.5:1 or slightly higher, but the engine's state of tune is extremely critical at this point, and there is very little margin for error.

Porting Advice

It's not always possible to find an as-cast cylinder head with the proper amount of airflow and port volume for an application. Many professional Pontiac engine builders can supply a pair of cylinder heads that are tailored for your specific engine.

If you plan to port your own cylinder heads, there are several points to keep in mind. Airflow follows fluid dynamics and doesn't like to change direction or compress or expand abruptly. The cross-sectional area of an intake port should be as consistent as possible. The port should taper gradually toward the valve bowl. The transition from the valve bowl into the valve seat, and then as it expands into the combustion chamber, should be smooth and consistent.

Porting an original cast-iron D-port is an easy way to increase airflow. Some hobbyists are willing to perform the task at home or pay for a professional service, but hand porting is not nearly as popular as it was in the past. There are too many high-flow aftermarket offerings at a reasonable cost available today.

When porting always remember that, generally, a larger port flows more air than a similar, but smaller port. You might assume that simply enlarging a port improves airflow, and that sometimes does result, but it's important the additional port size is located in areas of the port where it can actually enhance airflow. If any portion of the intake port is stagnant, or has an area where port volume doesn't have a positive effect on airflow, then a better-shaped, smaller-sized port can, and often does, flow a greater airflow more consistently.

Cylinder Head Options

Not long ago the only cylinder heads available for the Pontiac V-8 were cast-iron originals. The market has exploded with cast-aluminum offerings in recent years and it seems as if there's more than one option for a specific application today. The following is a buyer's guide detailing the various castings and the applications each is best suited for.

Cast-Iron Pontiac D-Port

When Pontiac developed its 1955 287-ci V-8, its cylinder head featured a compact intake port that displaced slightly more than 150 cc. The small exhaust port featured a D-shaped outlet, and the center two exhaust ports were siamesed, allowing for a relatively small exhaust manifold package.

As engine displacement grew in the years that followed, Pontiac revised the intake port shape slightly and increased valve sizes to improve overall airflow. Though actual port volume remained near 153 cc, measured airflow from a typical D-port casting with a 2.11-inch intake valve peaked around 210 cfm at 28 inches. The combination maintained strong low-end torque, but improved the top end performance from the larger mills of the era.

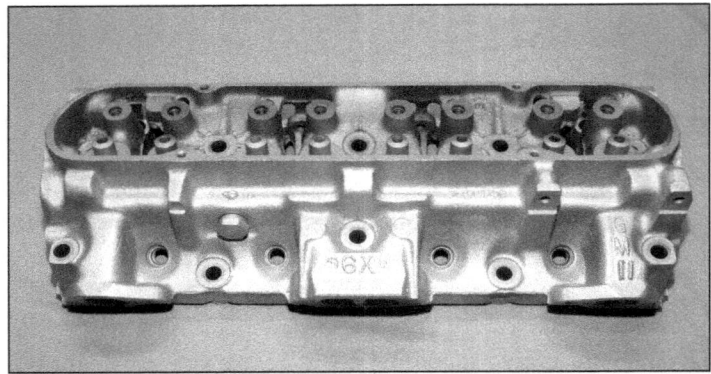

Most cast-iron Pontiac D-ports from 1968 forward are considered a viable option for performance use. The best feature 2.11/1.77-inch valves, threaded rocker arm studs, and six-bolt exhaust manifold flanges. Though fitted with 1.66-inch exhaust valves, the number-4X, -5C and -6X castings are some of the most affordable and commonly available.

Some cast-iron Pontiac D-ports were equipped with 1.96-inch intake valves. That limits peak intake airflow to about 190 cfm at 28 inches of pressure. Modifying the intake seat to accept a 2.11-inch valve is a simple process that most machine shops can perform for a couple hundred dollars. It's an easy way to increase peak airflow by about 20 cfm.

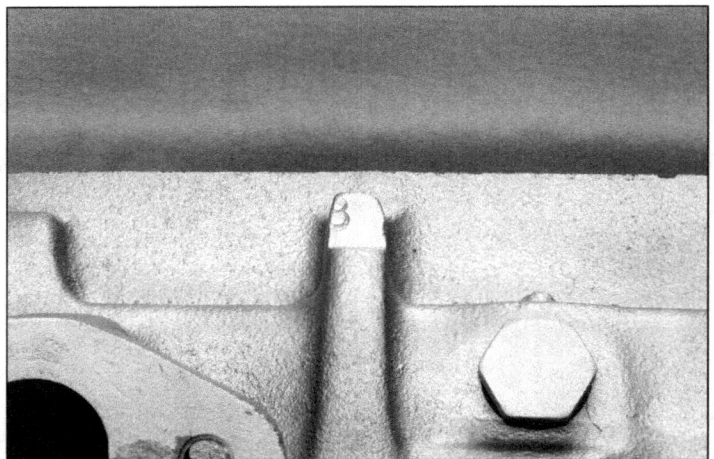

Pontiac began identifying its cylinder heads with a secondary application stamp in 1973. Found on the vertical accessory boss located between the left and center exhaust ports, the most desirable stamps are "4" and "8." A number-5C or -6X with a "4" has a 91- to 93-cc chamber while those with an "8" have a 100 cc chamber.

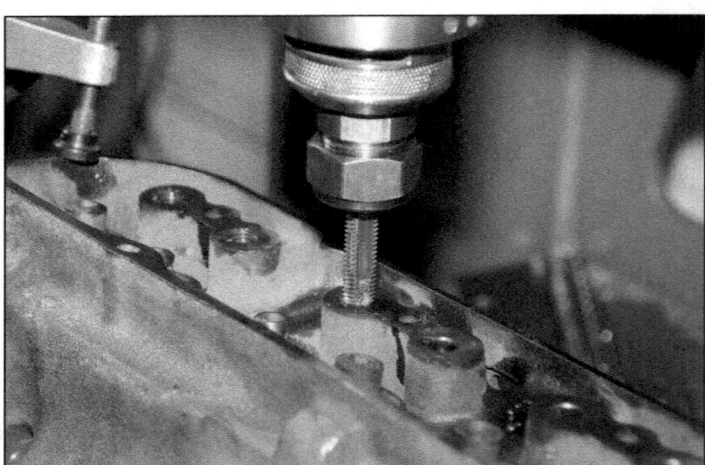

Pontiac commonly used pressed rocker studs on its low-performance applications. An aggressive camshaft can pull them out of the boss or cause the 3/8-inch bottleneck design to break off. A machine shop can pull the original studs and tap the bosses to accept oversized 7/16-inch rocker arm studs for about $100.

Original Pontiac D-port cylinder heads are not overly difficult to find today. The small-chamber castings associated with the late-1960s performance era are rather rare and quite costly. Castings from the mid- to late-1970s "smog era" were once considered useless simply due to negative perception of any performance component of that period. Contrarily, the number-4X, -5C, and -6X castings make excellent, affordable choices for high-performance rebuilds.

Low-performance applications from the late 1960s and early 1970s were fitted with smaller 1.96/1.66-inch valves and/or pressed rocker arm studs. Peak intake airflow typically measured around 190 cfm. Your machinist can easily modify an original small-valve casting to accept larger 2.11/1.77-inch valves for a reasonable cost. It will quickly and easily boost airflow toward 210 cfm, and additional porting can significantly increase it beyond that. Your machinist can also convert any original-pressed-stud casting to threaded studs for minimal cost.

The flow characteristic of the original D-ports are so similar that I suggest being less concerned with finding a specific casting number, and instead

CYLINDER HEADS

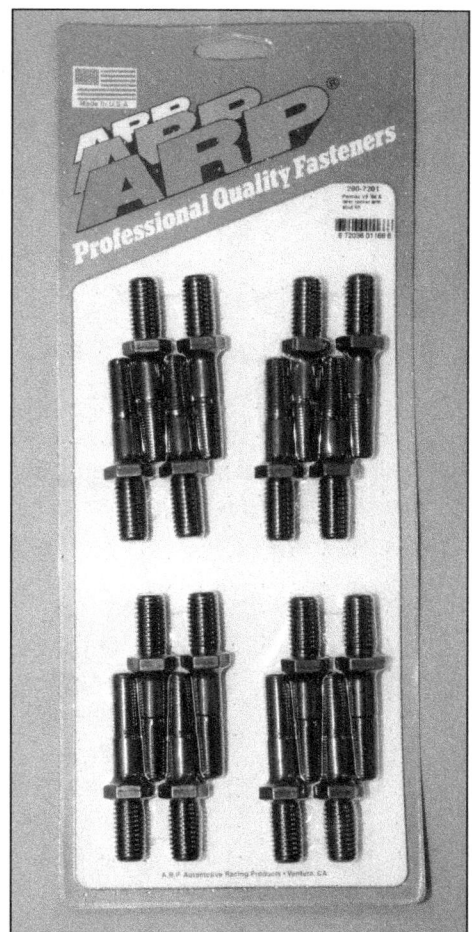

ARP number-290-7201 rocker stud kit is specifically designed for converting an original pressed-stud cylinder head to threaded studs. The ARP studs feature a typical 7/16-inch upper with a 1/2-20 base. The base threads are purposely left slightly long and must sometimes be trimmed to fit.

SD Performance has developed CNC porting programs for popular cast-iron Pontiac cylinder heads. Delivered fully machined and ready to install for a reasonable price, airflow increases toward 250 cfm at 28 inches in most cases. Optional porting takes that toward 280 cfm or slightly more. In my opinion, the CNC-ported cast-iron D-port is much more sensible and cost effective than attempting to port your own.

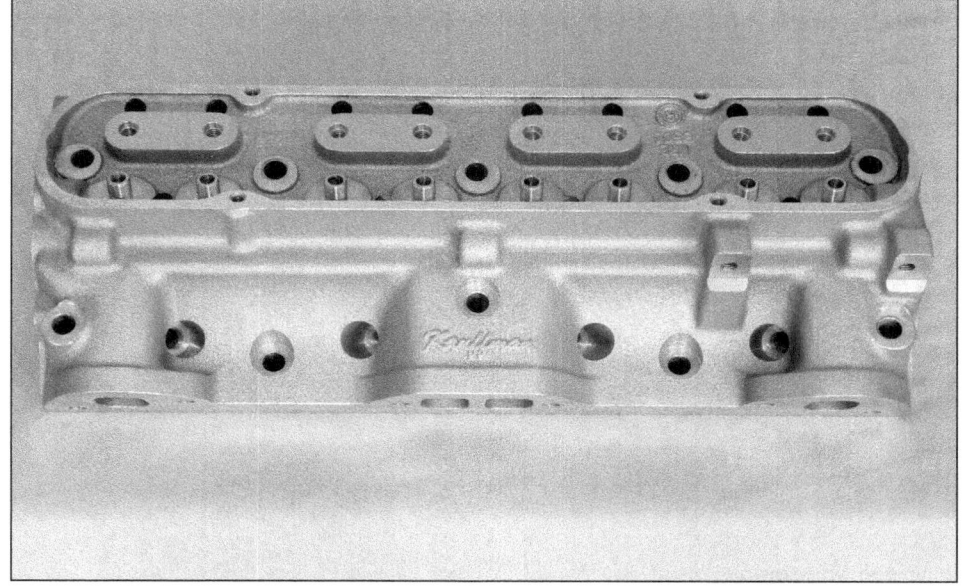

KRE developed a high-flow cast-aluminum D-port as a modern option for high-performance street engines. The castings have been on the market for several years. Available bare or fully assembled, it is a popular option with hobbyists.

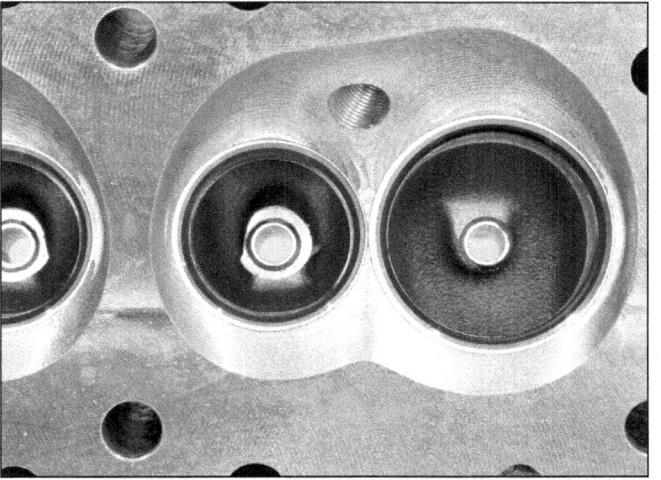

The KRE D-port includes a fast-burn combustion chamber that houses standard 2.11/1.66-inch valves. Offering more than 250 cfm of peak intake airflow and a small, compact intake port for strong port velocity, some companies are seeing as much as 340 cfm with aggressive porting and larger valve sizes.

search for a casting that has the exact amount or slightly more chamber volume than what's required to achieve your target compression ratio. Your machinist can easily mill the deck surface to reduce chamber volume or cut the valve seats for larger valves.

A typical D-port with a 2.11-inch intake valve can easily support just over 400 hp. Expect to spend several hundred dollars preparing a cast-iron D-port for a high-performance application; and that doesn't include surfacing, valves, hardened valve seats, and a high-quality valve job or bronze valveguides.

Your engine builder can properly prepare your cast-iron D-ports for you, but in my opinion, one of the best options is from SD Performance. It has developed a number of CNC-based porting programs that increase airflow to at least 250 cfm, and to as much as 280 cfm depending upon the casting for a very reasonable cost.

Cast-iron D-ports remain popular with hobbyists building high performance engines on a budget, those racing in classes where original castings are required, or those who are simply Pontiac purists. Beyond that, it's impractical to spend hundreds of dollars having a pair professionally ported and rebuilt. Any of the entry-level aftermarket Pontiac V-8 cast-aluminum cylinder heads flow nearly as much in as-cast condition and are ready to run, and can easily flow substantially more than a max-ported cast-iron D-port for a reasonable cost.

KRE D-Port

One of the most popular aftermarket Pontiac V-8 cylinder heads today is the cast-aluminum D-port developed by Kauffman Racing Equipment (KRE) in the early 2000s.

With 2.11/1.66-inch valves, peak intake airflow of the KRE D-port measures about 260 cfm at 28 inches of pressure and the exhaust flows more than 70 percent of that (about 185 cfm). Port velocity is an important factor in achieving strong low-speed performance for a street engine. KRE chose to limit intake port volume to 185 cc to promote good throttle response and high-torque output. Excellent peak flow allows it to consistently deliver 500 or more hp on moderate builds.

A key feature of the KRE D-port is its combustion chamber. The fast-burn design is similar to that of modern performance engines and promotes a quick

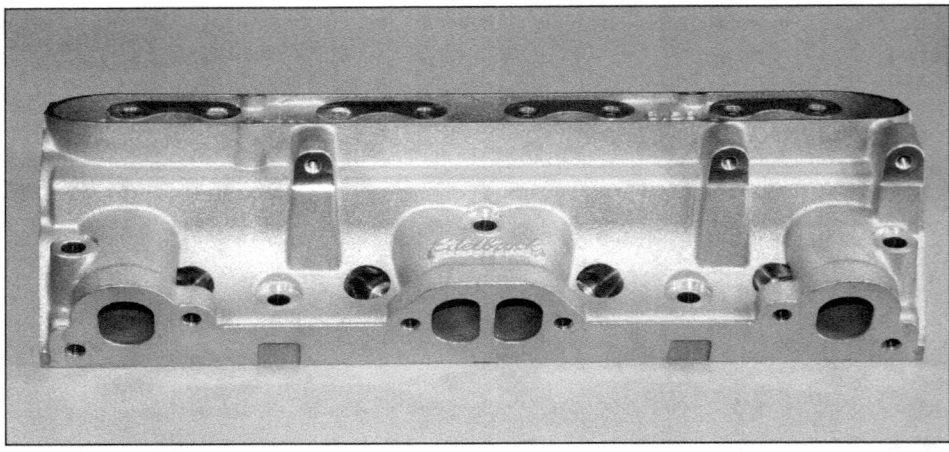

Edelbrock has recently released its own cast-aluminum D-port. As part of its Performer line, the modern interpretation includes an exhaust crossover and is 50-state emissions legal. It boasts of improved oil drainages for quicker return to the pan.

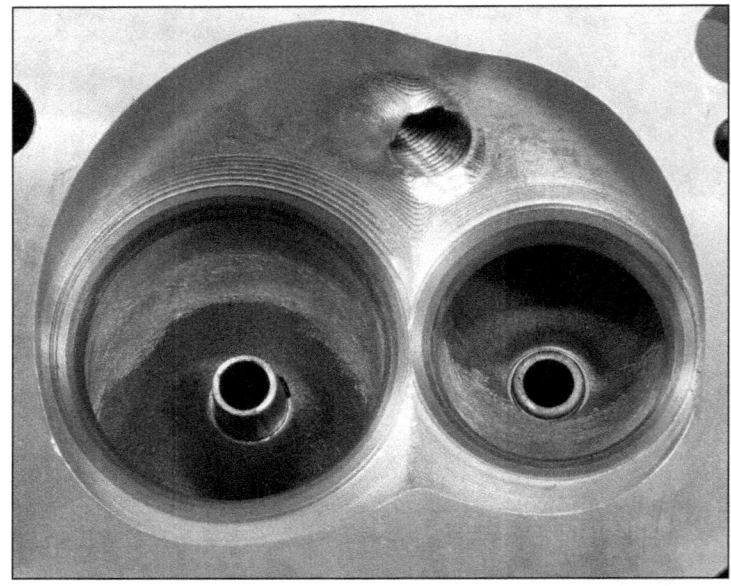

The Edelbrock Performer D-port uses 2.11/1.66-inch valves and airflow peaks at approximately 260 cfm at 28 inches. The intake and exhaust valve bowls are blended by hand for maximum consistency and performance. The fast-burn combustion chamber is available in several sizes.

and efficient burn. As-cast chamber volume measures 65 cc, but it can be CNC machined to 74 or 85 cc at an additional cost. Independent testing reveals that 30 and 32 degrees of total spark lead is all that's required for maximum performance. For comparison, a typical Pontiac 6X casting requires 36 to 38 degrees to achieve the same performance.

The KRE D-port is available fully machined, bare for professional finishing, or as complete castings that are ready to install. The latter includes stainless-steel valves and high-quality valvesprings, bronze valveguides with Teflon seals, and 7/16-inch ARP rocker studs. A pair of as-cast KRE D-ports sells for approximately $2,000.

Though the KRE D-port was originally designed as a replacement for high-performance street engines, it's proven its merit on street/strip and competition engines as well. Along with KRE, many Pontiac builders can boost peak airflow to 290 cfm and to as much as 340 cfm for a reasonable cost. Even at 340 cfm, they still bolt to any Pontiac V-8 block, use a typical Pontiac intake manifold, and do not require any offset rocker arms.

The KRE D-port is extremely popular with performance enthusiasts. I consider the as-cast KRE D-port an excellent alternative to any cast-iron original. It's much lighter and fits and installs just as an original. Overall airflow is substantially greater and it affords the distinct advantage of a modern fast-burn combustion chamber.

Edelbrock Performer D-Port

Edelbrock has produced its round-port replacement Performer RPM cylinder heads for several years, and it has been a very successful endeavor for the company. It recently ventured into the D-port market with the release of its cast-aluminum Performer.

Developed as its entry-level performance offering, Edelbrock's Performer D-port installs onto any Pontiac V-8 just like a cast-iron original. Key features of its design include high-airflow capacity, a modern combustion chamber, and vastly improved oil drainage passages. A cost of about $2,200 per pair is quite reasonable.

The Performer D-port includes 2.11/1.66-inch valves. The as-cast intake port volume measures 204 cc, and it's capable of delivering 260 cfm of peak intake airflow. The exhaust flows about 70 percent of that (182 cfm). After being on the market for only a few months, Pontiac professionals quickly increased peak intake airflow to 300 cfm, with the promise of 330 cfm or slightly more to follow.

Its heart-shaped combustion chamber is a modern, dual-quench design for maximum efficiency. It displaces 65 cc as cast, with 72- and 87-cc chambers available with CNC machining. Unlike the Performer RPM, where the larger chamber raises the valve seat and negatively affects the short turn transition, subsequently reducing airflow by several CFM at all lift points, Edelbrock increased the deck thickness of its 87-cc Performer D-port casting by .050 inch, which increases chamber volume without altering valve seat position.

The Performer D-port is an excellent alternative wherever a cast-iron D-port is a consideration, particularly any application where a peak intake airflow boost is required. A variety of combustion chamber volumes allows it to produce a good compression ratio on most Pontiacs.

Pontiac Cast-Iron Round Port

For Pontiac to remain competitive in the late-1960s' high-performance market,

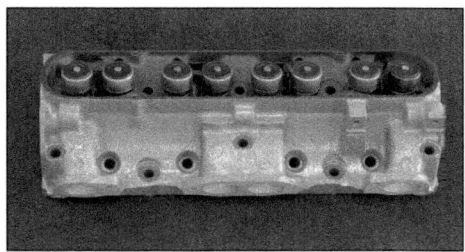

During the high-performance craze of the late 1960s, Pontiac developed a series of high-flow cylinder heads with round exhaust port outlets. These round-port castings were originally installed on such engines as the Ram Air IV and Super Duty 455. They boast of improved intake and exhaust airflow, and remain very rare and desirable.

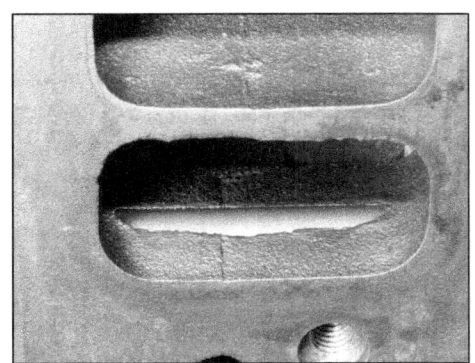

To maintain constant cross-sectional area within the SD-455 intake port, Pontiac widened it so much that the pushrod guide hole drilled into the casting broke through the sidewall of the intake port. A .030-inch sleeve was pressed into place.

it recognized the need for larger, more powerful engines capable of continuous high-speed operation. A series of new cylinder heads boasting of improved airflow were developed to substantiate strong top end performance. The castings were limited to the 1968½ Ram Air II, 1969–1970 Ram Air IV, 1971–72 455 H.O., and 1973–1974 Super Duty 455 engines.

While the number-96 R/A II casting contains an intake port that's nearly identical to the number-670 D-port, and subsequently flows around 210 cfm, the intake port volume of the R/A IV, 455 H.O., and Super Duty 455 was increased at least 170 cc, and even more depending upon the application. The larger port increases overall flow and cross-sectional area. Peak intake airflow measures around

The intake and exhaust ports of the Pontiac round ports were noticeably wider and larger than a typical D-port. The 1973–1974 Super Duty 455 casting (shown here) has a very smooth and contoured shape. Its intake port volume measures more than 185 cc and flows as much as 245 cfm at .550-inch lift.

CHAPTER 6

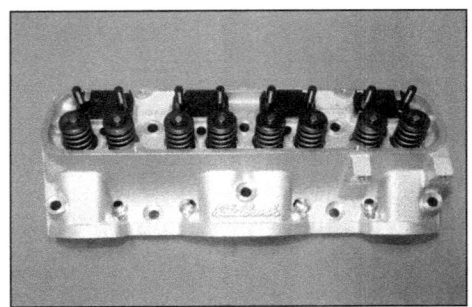

The cast-aluminum Performer RPM from Edelbrock is widely used by performance enthusiasts. Introduced in the 1990s as a high-flow replacement for the valuable Pontiac round-ports, its popularity and effectiveness hasn't waned. It's available semi-finished, fully finished but bare, and fully assembled and ready for installation.

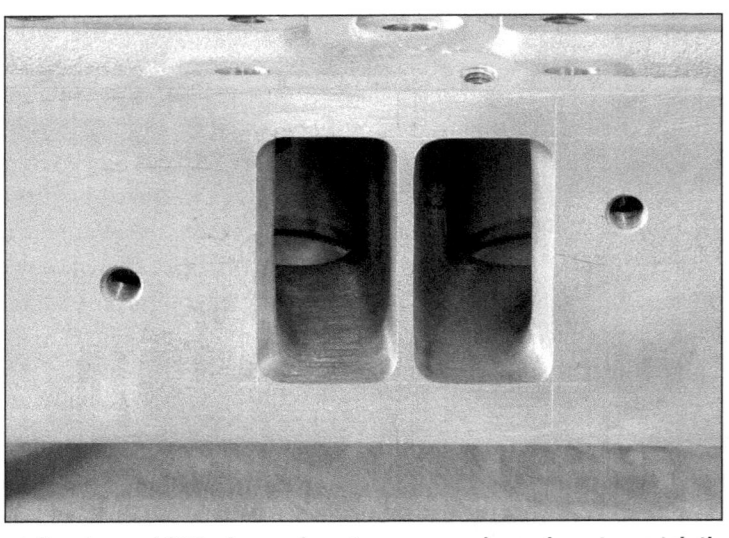

Many professional Pontiac engine builders offer a blueprinting service for the Edelbrock RPM at a very reasonable cost. It includes gasket matching the intake port, a 1.77-inch exhaust valve, smoothing and blending the intake and exhaust ports taking peak airflow toward 300 cfm, and custom-spec valvesprings to match the camshaft. For the couple hundred dollars it adds to the cost of an otherwise complete casting, it produces one of the best performance values available today.

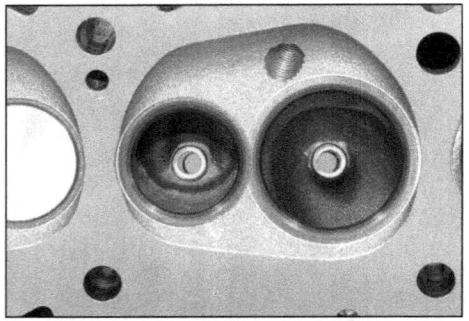

The Performer RPM uses a wedge-type combustion chamber available in with 72 or 87 cc. Less efficient than a fast-burn chamber, 36 to 40 degrees of total spark lead is commonly required to achieve peak power. Standard valve sizes are 2.11/1.66 inches and airflow peaks at about 280 cfm in as-cast form. Edelbrock hand blends the intake and exhaust bowl seats for improved airflow and consistency.

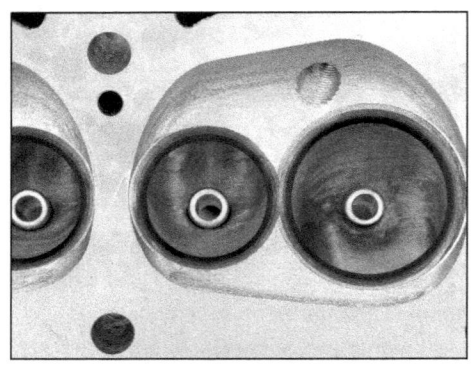

For engines that require additional airflow to reach a target horsepower peak, the Performer RPM can be modified for even greater flow potential. With a combination of a 2.19-inch intake valve and additional porting, many Pontiac builders offer Edelbrock Performer RPMs that peak at a minimum 330 cfm of intake airflow, and possibly slightly more.

240 cfm at .500- to .550-inch valve lift from the R/A IV and SD-455 castings, while the smaller 455 H.O. intake port peaks at slightly less.

The round-port castings were coveted by racers for years, particularly the 1968½ R/A II number-96s, and the numbers-722 and -614 from the 1969–1970 R/A IV because of their good flow characteristics and small combustion chambers.

A virgin pair of round-port cylinder heads can still bring several hundred to several thousand dollars depending upon the application and date code. Still able to provide stout performance, considering the rarity and expense of such castings, an original round-port is far more valuable to restorers than racers in today's market.

Edelbrock Performer RPM

Edelbrock responded to the hobbyist need for a high-flowing Pontiac V-8 cylinder head during the mid 1990s when it began producing its cast-aluminum Performer RPM. Over the years, the Performer RPM has proven itself as an excellent choice for high-performance street and race engines alike.

The Performer RPM offers 280 cfm of peak intake airflow while exhaust peaks at nearly 70 percent of that (about 196 cfm). Edelbrock offers two distinct combustion chamber volumes: 72 or 87 cc. The larger 87-cc chamber relocates the valve seat and positions it higher in relation to the port floor. When compared to the 72-cc casting, airflow is negatively affected. The deficit is relatively minor, however, measuring no more than 10 cfm at any lift point.

Intake port volume measures 215 cc. When compared to an original Pontiac D-port, that may seem excessively large. When you consider that most modern builds include at least a 460-ci engine, the cast port volume is adequate for strong peak horsepower, and nearly ideal for the larger 500-plus-inch engines that are so common today.

CYLINDER HEADS

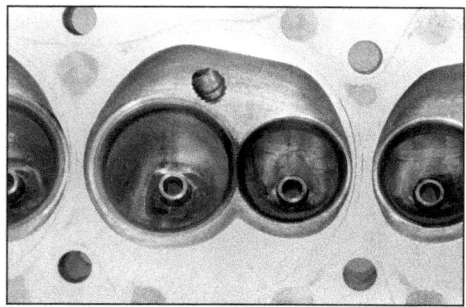

Edelbrock says that it may someday update its Performer RPM, adding a modern, fast-burn combustion chamber. In the meantime SD Performance has improved its combustion efficiency by adding material to the as-cast 72-cc combustion chamber and CNC machining a heart-shaped appearance. The result is a slight power increase with less total spark lead. The option costs a few hundred dollars.

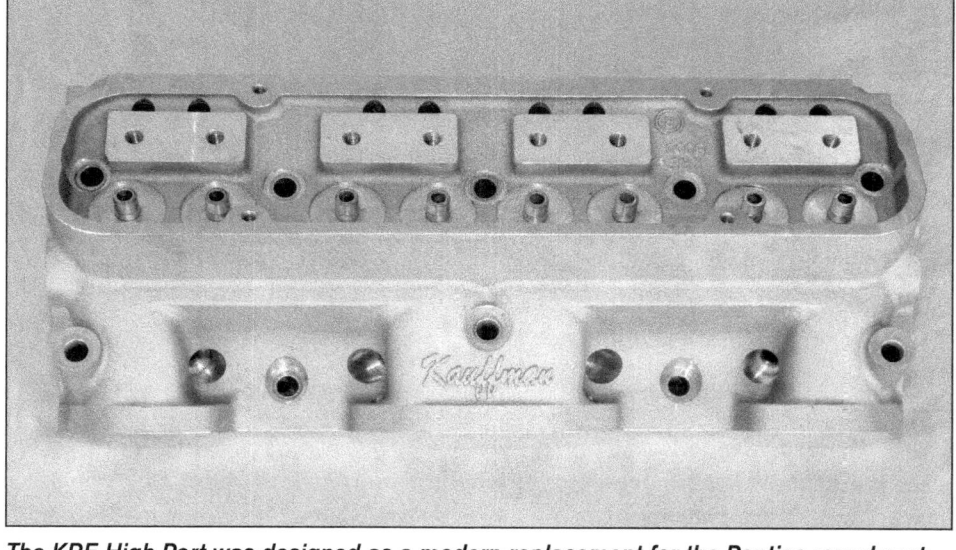

The KRE High Port was designed as a modern replacement for the Pontiac round port. Peak intake airflow measures 330 cfm as cast, but optional porting increases that toward 400 cfm. It uses a typical intake manifold and round-port exhaust headers.

Butler Performance and SD Performance maximize Performer RPM intake airflow with a "wide port" option (right). It peaks at 370 cfm or slightly more. Intake port volume is substantially greater. Intake manifold choice is limited to a few aftermarket units or a custom-fabbed sheet-metal unit. Exhaust port configuration is unchanged, and remains a typical Pontiac round-port.

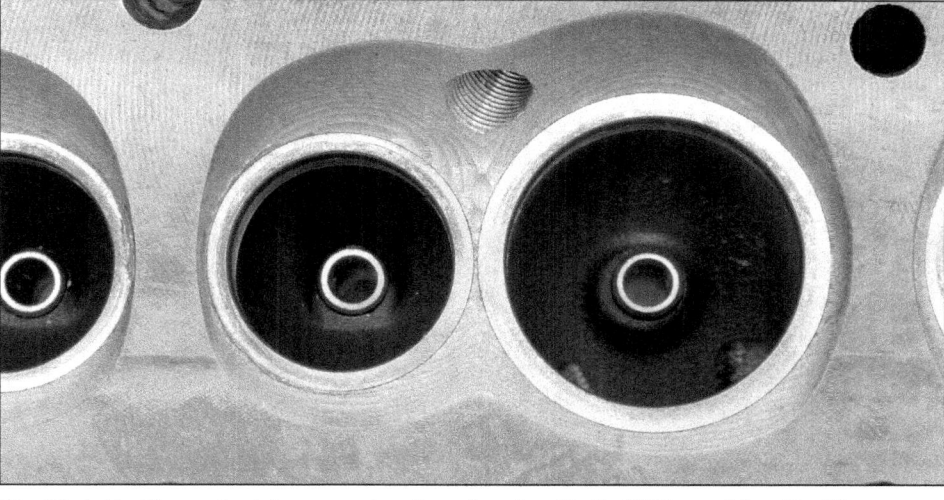

The High Port has a fast-burn combustion chamber that's CNC machined to 56 cc. Other chamber volume options include 64 and 80 cc. Standard valve sizes measure 2.20/1.70 inches with 2.25/1.75-inch valves optional.

Edelbrock's Performer RPM is available semi-finished, fully machined but bare, and as complete castings ready to install. The semi-finished offering requires professional finishing. The complete casting includes stainless-steel 2.11/1.66-inch valves, a single valvespring package that's intended for a flat-tappet camshaft with as much as .575-inch valve lift, bronze valveguides, and 7/16-inch ARP rocker studs. Complete Performer RPM castings sell for approximately $2,000 per pair. That's an excellent value for a high-quality cylinder head that easily supports 500 hp out of the box and offers plenty of potential for future modifications.

The Performer RPM is quite capable in as-cast form, but professional engine builders have exploited its versatility for engines requiring substantially more airflow. For a performance goal greater than 500 hp, minor port cleanup that increases peak intake airflow to at least 300 cfm is available from most Pontiac builders for a few hundred dollars. The best options are those where intake port volume is marginally increased, maintaining strong port velocity and low-lift airflow. Exhaust flow is typically 70 to 75 percent of the intake.

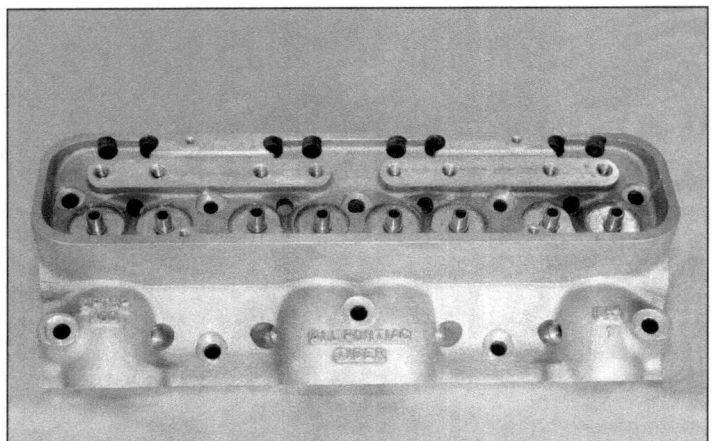 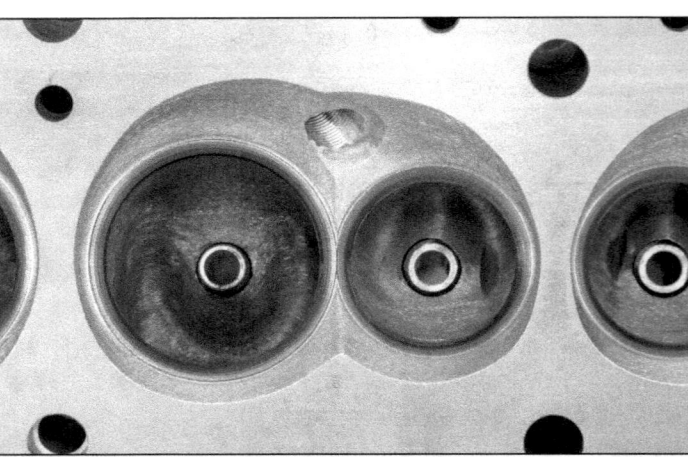

The Tiger head from AllPontiac.com is popular with racers running the quickest. Peak intake airflow is greater than 410 cfm as cast. It accepts a traditional Pontiac intake manifold, but requires an intake flange adapter when using a typical intake manifold.

The Tiger head features a fast-burn chamber that measures 66 cc. No other optional volumes are available. AllPontiac.com recommends using dished pistons to achieve the target compression ratio. Standard valve sizes measure 2.25/1.80 inches.

Where the performance goal is 600 to 700 hp, peak intake airflow increases up to 330 cfm or slightly more are commonly available. Many Pontiac builders have developed complete Performer RPM packages that include additional machining and porting, oversized valves, specific valvesprings, and lightweight valvetrain components for a reasonable cost.

Butler Performance and SD Performance have developed a highly specialized Performer RPM for those seeking to produce as much as 900 hp or slightly more. The "wide-port" design features a large intake port that measures about 275 cc. Intake airflow peaks near 375 cfm, but the port is so wide at the entrance that the intake pushrod must be relocated. That requires a lifter with an offset pushrod cup, a rocker shaft system with intake rocker arms that are offset 1/2 inch or more, or both. Expect to spend at least $3,500 for a pair of wide-port castings plus the associated required components.

KRE High Port

Not long after KRE introduced its cast-aluminum D-port, it developed an aluminum casting for larger-displacement engines. The High Port is a high-flow replacement that bolts to any conventional Pontiac V-8 block, uses the stock valvetrain configuration, and accepts any standard Pontiac intake manifold and round-port exhaust header.

Designed for a minimum bore diameter of 4.15 inches, the High Port uses 2.20/1.70-inch valves. The as-cast intake port volume measures 277 cc and airflow peaks at 330 cfm. The exhaust flows about 70 percent of that (about 230 cfm). Intake airflow of as much as 430 cfm is possible with additional porting and 2.25/1.75-inch valves, but it requires the use of a rocker shaft system with intake rocker arms that are offset at least .700 inch.

The High Port combustion chamber is a modern fast-burn design. It's CNC-machined and features a distinct heart-shaped appearance. Standard volume measures 56 cc, but various volumes up to 80 cc are available at additional cost. KRE offers its High Port bare, but with valve seats, bronze guides, and complete castings that include stainless-steel valves, valvesprings compatible with a solid-roller camshaft, heavy-duty valvetrain hardware, and 7/16-inch ARP rocker studs. A pair of complete castings begins at $2,500.

The KRE D-port may be a better choice for a moderately sized street engine, or even one that may see the drag strip occasionally. The High Port is ideal for max-performance engines displacing at least 467 ci and where 625 hp or greater is the goal. With additional port work and the right combination of components, the High Port can sustain around 900 hp.

AllPontiac.com Tiger

Several years ago DCI Motorsports began developing a high-flow cylinder head that would allow large-cube Pontiac V-8s to compete with similarly sized big-block Chevys on the drag strip. When the project stalled, AllPontiac.com purchased the rights and tooling to the new Tiger head, and saw the aluminum casting into production during the mid 2000s.

The as-cast intake port volume measures 310 cc, and it peaks at 414 cfm of intake airflow. The exhaust port flows more than 70 percent of that (290 cfm). The Tiger head is available as a bare casting fitted with valve seats and bronze valveguides (but requires professional finishing) and a complete casting that's ready to install. The bolt-on package includes 2.25/1.80-inch valves and a

CYLINDER HEADS

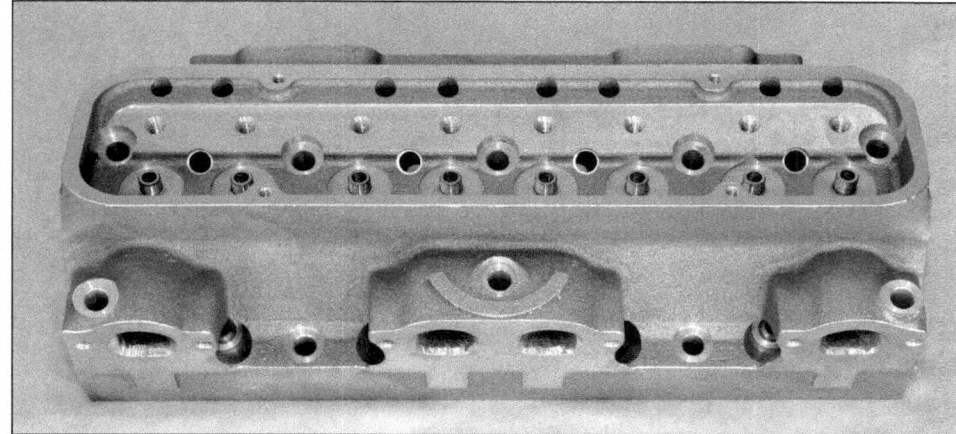

The Wenzler Super Chief is a semi-finished head that requires professional finishing for its intended application. Intake airflow peaks around 320 cfm as cast, but as much as 430 cfm is possible with proper porting. The intake ports are so wide that a rocker shaft system is required.

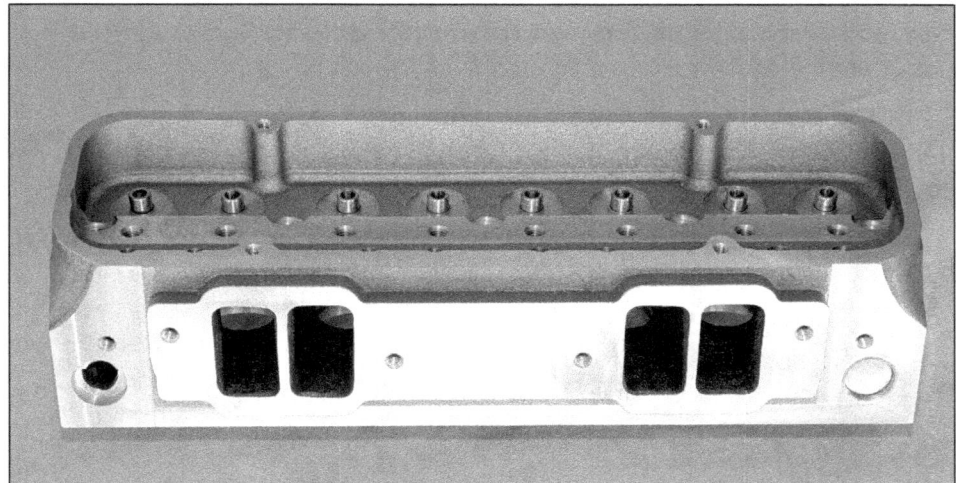

The Super Chief has raised intake runners to improve airflow. The intake flange features a cast runner extension to allow the use of typical Pontiac intake manifold. It's machined off when the Wenzler Gutsram intake manifold is used.

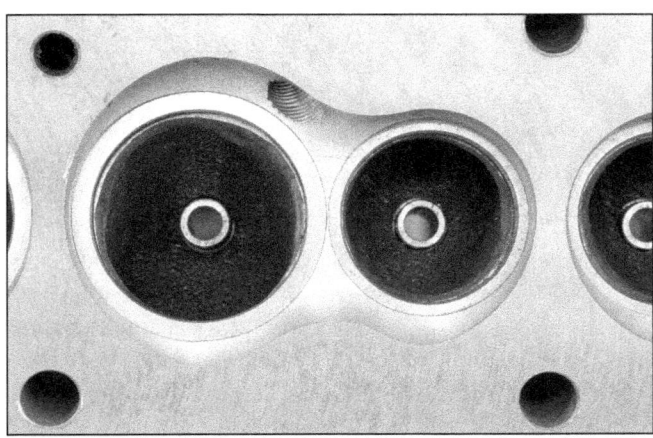

The Super Chief has a fast-burn combustion chamber that can be CNC machined to most volumes between 56 and 85 cc. You must specify the desired volume at the time of ordering. Otherwise your Pontiac professional can hand finish it for you.

number of valvetrain options for moderate- to high-lift roller camshafts.

Though the Tiger head uses a typical Pontiac intake manifold arrangement, it has a much taller deck height, and that alters the point at which the intake manifold and cylinder head meet. AllPontiac.com offers its own 4-barrel Tiger intake manifold with extended runner length, or intake flange adapters to allow the use of a conventional aftermarket Pontiac manifold. Two distinct exhaust port configurations are available. One accepts a typical round-port tube header and the other features a raised-port design that requires a custom tube header.

The Tiger head features a fast-burn combustion chamber that measures 50 cc in as-cast form. The chamber of the complete casting is CNC machined and displaces 64 cc. Requiring a minimum bore diameter of 4.31 inches, the Tiger head is capable of delivering more than 900 hp on a typical 500-inch Pontiac V-8. Some of the fastest and most powerful Pontiac-powered cars on the drag strip today are equipped with Tiger heads. Selling for about $6,000 per pair ready to install, they're among the best performance value for your Pontiac V-8.

Wenzler Super Chief

Wenzler Engineering was among the first aftermarket companies to produce a cast-aluminum Pontiac V-8 cylinder head when it introduced its original round-port during the 1980s. Though still available as the "Series II," the Super Chief is Wenzler's high-flow offering specifically developed to extract maximum potential from a max-performance Pontiac V-8.

The Super Chief is available only in a semi-finished state. It is CNC machined and delivered with valve seats and bronze valveguides installed. The as-cast intake port flows nearly 320 cfm of peak intake airflow, and with professional porting, the casting can deliver as much as 430 cfm, and possibly slightly more. The as-cast exhaust port airflow is about 70 percent of the intake port (about 225 cfm), and responds similarly to professional porting.

Wenzler recommends using 2.25/1.80-inch valves when finishing the Super Chief. The intake bowl contains enough material to enlarge the intake valve to 2.30 inches for greater flow. Valve shrouding can be an issue when combining the recommended valve package with a bore diameter less than 4.25 inches, so Wenzler recommends at least 4.375 inches, but less may certainly be possible. The Super Chief also requires a rocker shaft system with offset intake rocker arms to accommodate the wide-diameter intake ports.

The Super Chief is cast with intake runner extensions that allow the use of any conventional aftermarket Pontiac intake manifold. Wenzler also produces its Gutsram intake manifold specifically for the Super Chief. The exhaust port uses a typical round-port flange, but because its outlets are raised to improve airflow, a custom tubular header is required.

Wenlzer is a small company producing specialized cylinder heads for certain applications. It doesn't advertise and may not be as well known as some of other larger companies producing aftermarket cylinder heads, but the Super Chief has proven itself as an excellent choice for serious performance. It has delivered nearly 1,100 hp when properly modified and offers excellent performance potential.

Edelbrock Pro Port Raw

Edelbrock's Performer RPM was developed as a high-performance replacement for the original Pontiac round-ports, and significant airflow increases are possible with extensive modifications. Builders have maximized its flow limit, however. Edelbrock's new Victor-series Pro Port Raw is its answer for the next level of

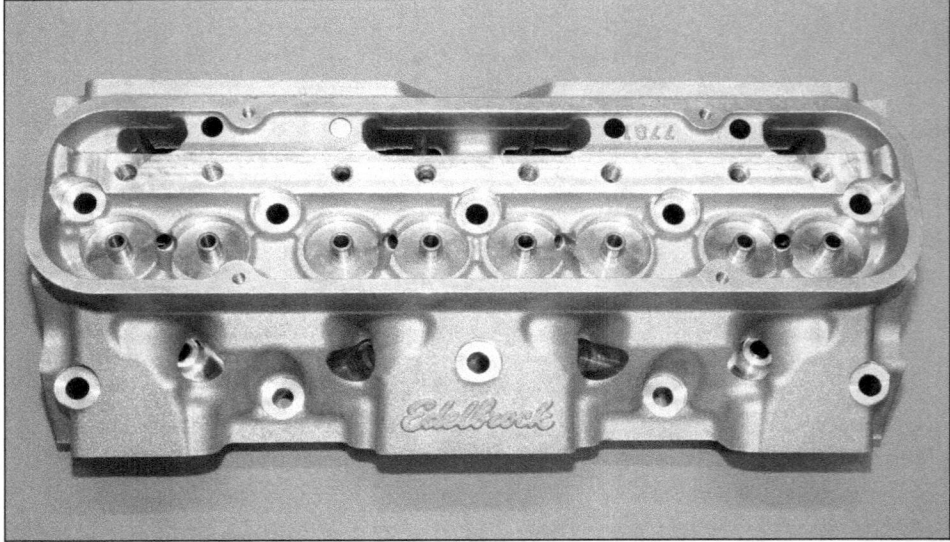

The Edelbrock Victor-series Pro Port Raw is a new and exciting cylinder head for max-performance Pontiac V-8s. It's delivered minimally machined and must be completely finished for any application. Two distinct castings are available. One has a valve-to-piston angle of 12.6 degrees and the other is 11 degrees.

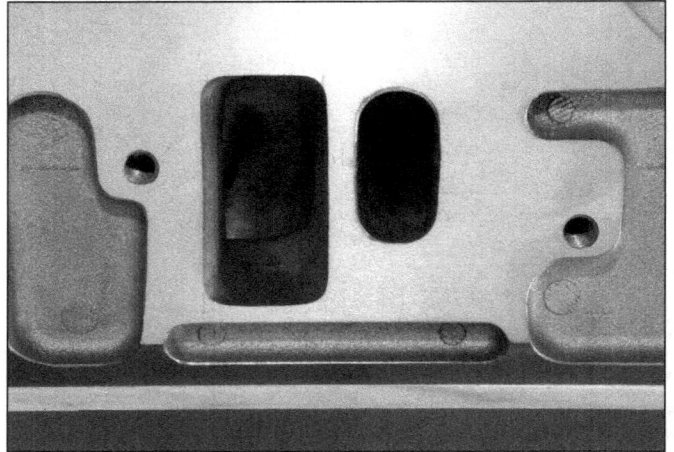

The Pro Port Raw intake and exhaust ports are very small and completely unfinished (right). They must be professionally ported to the desired size and level of airflow (left). Intake airflow of 400 cfm is possible with either. The 11-degree casting accommodates 475 cfm or slightly more.

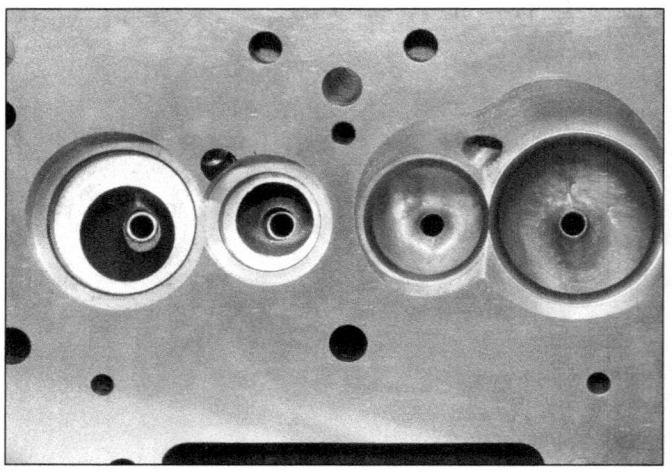

Like its intake and exhaust ports, the Pro Port Raw combustion chamber is also completely unfinished. The spark plug is pointed toward the exhaust valve for maximum efficiency. It can be finished with a heart-shaped or wedge-type chamber and any volume.

Pontiac performance, and incorporates many design features for the future performance enhancements.

When you're a serious racer, cylinder heads almost always require some type of custom modifications and the Pro Port Raw is delivered semi-machined. In as-cast form, its intake and exhaust ports are very small and completely unfinished. They must be ported by a professional to the desired shape and amount of airflow. Its combustion chamber is also unfinished and must be finished to the desired shape and size.

Edelbrock offers two distinct Pro Port Raw castings; valve angle from the piston centerline is the difference between them. Reducing the valve angle influences airflow and certain racing bodies allow no more than a 2-degree variance from stock. For Pontiacs built from 1967 forward, the stock angle is 14 degrees and Edelbrock offers a Pro Port Raw with a valve angle of 12.6 degrees. The other casting features a valve angle of 11 degrees for those racing unrestricted.

To improve airflow, Edelbrock lengthened the intake port of the 12.6-degree casting by 5/8 inch. The intake port of the 11-degree casting is lengthened 1.25 inches, which lends a tunnel port effect, significantly improving airflow and performance. With typical port work, Edelbrock reports that the 12.6-degree Pro Port Raw intake port can deliver at least 360 cfm with 2.19/1.74-inch valves and it has seen with as much as 420 cfm. At least 400 cfm is available from the 11-degree casting with as much as 475 cfm possible with additional porting.

Even with the added intake port length, Edelbrock specifically designed its Pro Port Raw to accept a traditional Pontiac intake manifold. Depending upon the amount of porting required to attain the desired amount of airflow, the Pro Port Raw uses a typical pushrod location and stud-mounted rocker arms.

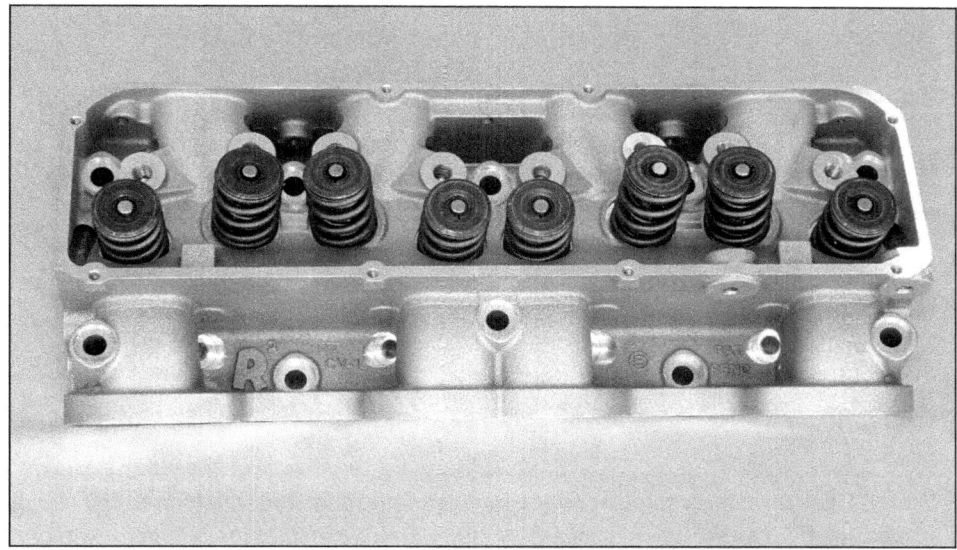

Roland Racing CV-1 features canted-valve angles. It requires a specific intake manifold, but uses a normal Pontiac camshaft layout and doesn't require offset intake pushrods. The exhaust port uses a typical round-port header, and a common big-block Chevy valve cover or others from Roland Racing.

Offset lifters and a rocker shaft system is required for maximum airflow efforts, however. The exhaust uses typical round-port tube headers.

The Pro Port Raw is for serious racing applications that require maximum airflow from the cylinder head package that uses a typical intake port configuration and valvetrain configuration. As of this writing, distribution from Edelbrock is limited, but with a suggested minimum bore diameter of 4.18 inches, it is sure to be an excellent choice for most engines of 460 ci or greater, and should give stunning results.

Roland Racing CV-1

When developing a high-flow aftermarket cylinder head, the diameter of the intake and exhaust valve package is limited by bore size. The closer the valves are to the cylinder walls, the more shrouding that occurs, which can negatively impact airflow. Canted-valve angles, where the intake and exhaust valves slightly oppose and point toward the cylinder centerline, allows for larger valves while lessening the risk of shrouding, maximizing airflow and performance.

Chevy used canted-valve technology when developing its big-block, and there are a great number of aftermarket companies producing canted-valve cylinder heads for many makes. It wasn't until Roland Racing released its cast-aluminum CV-1 in the 2000s that canted-valve technology was available to Pontiac hobbyists.

The CV-1 intake port measures 285 cc and is capable of flowing 380 cfm as cast. The exhaust port flows about 70 percent of the intake (about 265 cfm). As delivered, it uses 2.25/1.66-inch valves, but the CV-1 has extra material cast in for professional porting and additional airflow. It can easily accommodate 2.30/1.71-inch valves and a peak intake airflow increase to 440 cfm. Complete castings, which include quality stainless-steel valves and conical valvesprings, sell for $3,000.

As the first canted valve cylinder head developed for Pontiac V-8s available on a mass scale, the CV-1 easily bolts to any traditional or aftermarket

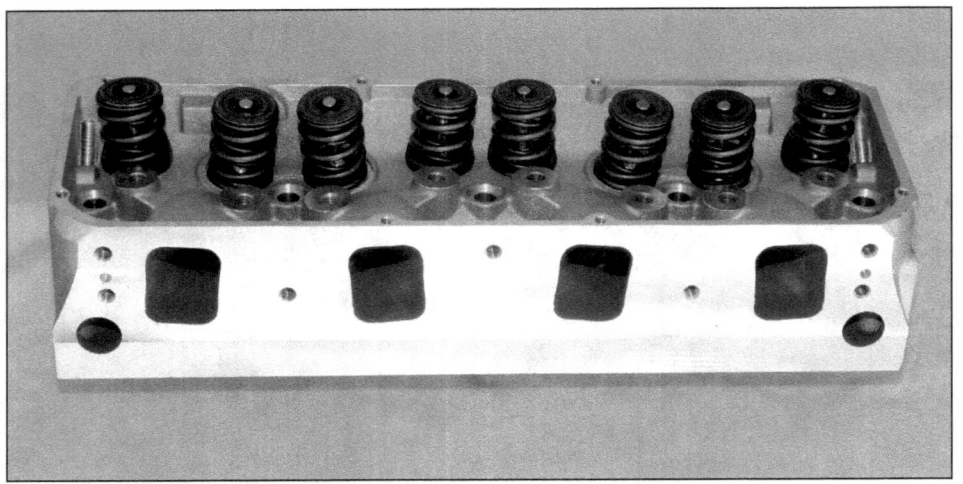

The CV-1 has unique rectangular intake ports that measure 285 cc and flow 380 cfm as cast. Ideal for larger engines, plenty of material was cast into the cylinder head for professional porting. Roland Racing offers a second CV-1 with a 215-cc intake port for smaller engines, or for those that want their CV-1s professionally finished elsewhere.

The CV-1 features a CNC-machined, fast-burn combustion chamber that measures 65, 81, or 92 cc. Standard valve sizes measure 2.25/1.66 inches. Special pistons are required for proper piston-to-valve clearance.

Pontiac V-8 block with a bore diameter of at least 4.15 inches. Its canted-valve design requires a unique intake manifold arrangement, and Roland Racing offers several proprietary options at extra cost. The exhaust port uses a typical round-port exhaust header, but the port can be milled away for applications that require the additional clearance.

It has taken a while for its popularity to catch on, but with some noticeable success on the dyno and drag strip, the CV-1 has gained notoriety and proven itself as a viable option for any max-performance Pontiac V-8. The relatively large intake port and high airflow tend to produce broad power from 4,000 rpm toward 7,500, with horsepower generally peaking around 7,000 rpm depending upon the application. A canted-valve cylinder head may not be for every enthusiast, but it's certainly an option if maximum performance is the goal.

KRE Warp-6

With its successful cast-aluminum D-port and High Port cylinder heads already on the market, KRE has recently introduced its new canted-valve Warp-6 casting. At the time of this writing, it's so new that there hasn't been a great deal of independent testing performed, but KRE's internal testing indicates that it's very capable. More than 1,100 hp was achieved using as-cast Warp-6 heads on a 535 inch on KRE's dyno.

A key component in its astounding performance is its symmetrical valve arrangement. Unlike a traditional Pontiac cylinder head, where intake runners share a common wall, the center exhaust ports are siamesed, and some intake and exhaust ports flow better than others, the Warp-6 ports are located in the same relative position regardless of the cylinder and each intake and exhaust port offers the same level of airflow. The design advantage improves cylinder-to-cylinder consistency and overall engine performance.

The Warp-6 casting is available fully machined, but bare, and completely finished and ready to install. The casting is designed for standard 2.25/1.75-inch valves. Intake port volume measures 399 cc and offers 460 cfm of peak intake airflow as cast. The exhaust port flows about 70 percent of that (about 320 cfm).

It can accommodate up to 2.40/1.85-inch units and at least 515 cfm with professional porting. Complete castings include titanium valves, specific springs for .900 inch of roller cam valvespring lift, and other quality valvetrain components. The fast-burn combustion chamber is CNC machined and measures 60 cc.

In addition to flow balance, the symmetrical valve arrangement also eliminates the siamesed exhaust valves from the center cylinders of each bank. Typically an area of extreme heat, particularly on high-end race engines, the heat can comprise the gaskets ability to maintain proper cylinder seal, ultimately leading to complete gasket failure. The symmetrical arrangement allows heat to dissipate evenly across the casting, improving gasket life.

Because of its unique valve arrangement, the Warp-6 requires a specific camshaft with proper lobe placement

CYLINDER HEADS

and a complete rocker shaft system that KRE developed in conjunction with T&D Machining. The exhaust ports require custom tube headers that measure more than 2 inches in diameter. KRE also offers specific intake manifolds, valve covers, and a coolant crossover water manifold, all of which are required for the Warp-6.

Unfortunately, you cannot simply install the Warp-6 casting and all its related components on to an existing Pontiac V-8 without major disassembly. The Warp-6 requires custom pistons with valve pockets strategically cut into the crown for proper clearance, and KRE can provide them for you. It may be best to dedicate a specific build to the Warp-6 if you plan to use them on your project. It looks to be a casting we can expect much from down the road.

Reproduction Ram Air V

Pontiac developed tunnel port cylinder heads for a group of new high-performance engines displacing 303, 400, and 428 ci. The 400-ci Ram Air V was very close to reaching production in 1969 before the tunnel port program was scrapped entirely. Pontiac estimates that about 200 pairs of 400-ci cylinder heads were produced at the time, and they remain very rare and quite valuable.

Hedging on the R/A V mystique, McCarty Racing developed a new tunnel-port Pontiac casting constructed of cast-aluminum that bolts to any stock or aftermarket Pontiac block but features a number of design improvements. At the time of this writing, McCarty Racing is finishing up its plan for complete bolt-on assemblies, so the final airflow numbers and valve size specifications are not yet finalized. Expect pricing to begin around $2,500 per pair and many options are available. A specific intake manifold and headers are required, both of which McCarty can provide at extra cost. It's best to contact the manufacturer for further details.

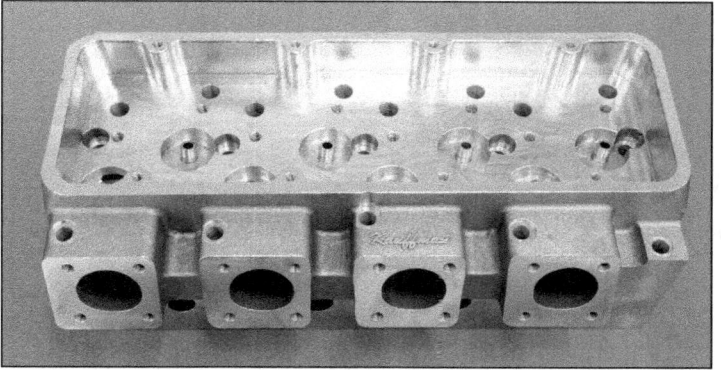

The KRE Warp-6 is the most recent offering in the canted-valve market for Pontiac V-8s. Its symmetrical valve configuration and port design provides maximum port consistency for greater performance. The valve arrangement reduces the excessive heat that builds between the siamesed exhaust valves on traditional castings that cause head gasket failure.

The Warp-6 fast-burn combustion chamber measures 60 cc. Peak intake airflow measures 460 cfm with standard 2.25/1.75-inch valves. That can increase to at least 515 cfm with optional porting and 2.40/1.85-inch valves. All associated components such as intake manifolds, header flanges, and valve covers, are available from KRE.

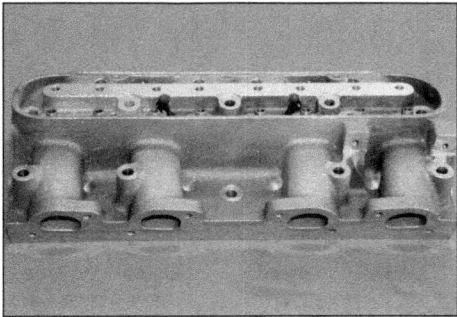

McCarty Racing has recently introduced its reproduction Ram Air V casting. Like the original, the reproduction features individual exhaust ports that require custom-made tube headers. It's compatible with any 400 or 455-ci Pontiac block and most any aftermarket unit. Specific fasteners are required and McCarty Racing can provide them.

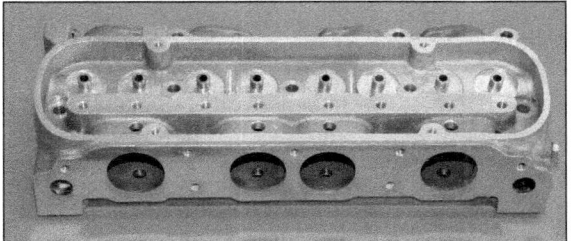

The Ram Air V was a tunnel-port engine that Pontiac developed during the late 1960s. It features large, round intake ports that have the pushrod running directly through the center. McCarty Racing's cast-aluminum reproduction mimics the original in nearly every respect, but the intake and exhaust ports are designed for improved airflow and the casting is generally beefier throughout. The pushrod's pressed guide tubes are missing from this preproduction example.

CHAPTER 7

VALVETRAIN

The camshaft opens and closes the valves at specific points in crankshaft rotation. Nearly all production engines were fitted with hydraulic valve lifters. Notice the "00" cast into the lifter galley. It denotes 400-ci displacement.

The Pontiac V-8 valvetrain is a simple overhead-valve design that's comprised of several different components. Each cylinder uses one intake and one exhaust valve that allows the engine to ingest and expel the combustible air/fuel mixture and spent exhaust gasses. A single camshaft located in the center of the block actuates the valves by transferring lobe lift through a .842-inch-diameter tappet (or lifter), a long pushrod, and a rocker arm that pivots on a ball socket and stud arrangement. The entire system is a very efficient and effective design.

Pontiac planned for its V-8 to use hydraulic lifters. During development the block's lifter bores were designed to deliver an adequate volume of pressurized oil for proper hydraulic lifter operation. The oil allows the lifter to continually adjust valve lash so that the entire valvetrain remains in constant contact, which provides a low-maintenance design that operates quietly and consistently throughout the life of the engine. With the exception of certain limited-production Super Duty and Ram Air engines that used mechanical camshafts for maximum performance, all other Pontiac V-8s were fitted with hydraulic lifters.

Hydraulic Lifters

A typical hydraulic lifter is quite complex. It consists of an outer body and internal plunger. Plunger depth within the body is controlled by a combination of oil and spring pressure. When the lifter is on the base lobe of the camshaft, pressurized oil enters the lifter body though a feed hole and floods a cavity located just beneath the plunger. As the cam rotates and the lobe lifts the lifter body, a spring-loaded check-ball within the plunger reacts to valvespring pressure, isolating the lifter from the engine oil supply. The lifter and plunger then raise as a unit, ultimately lifting the valve off its seat.

As the camshaft rotates, the oil trapped beneath the plunger is pressurized and bleeds outward as the plunger further

VALVETRAIN

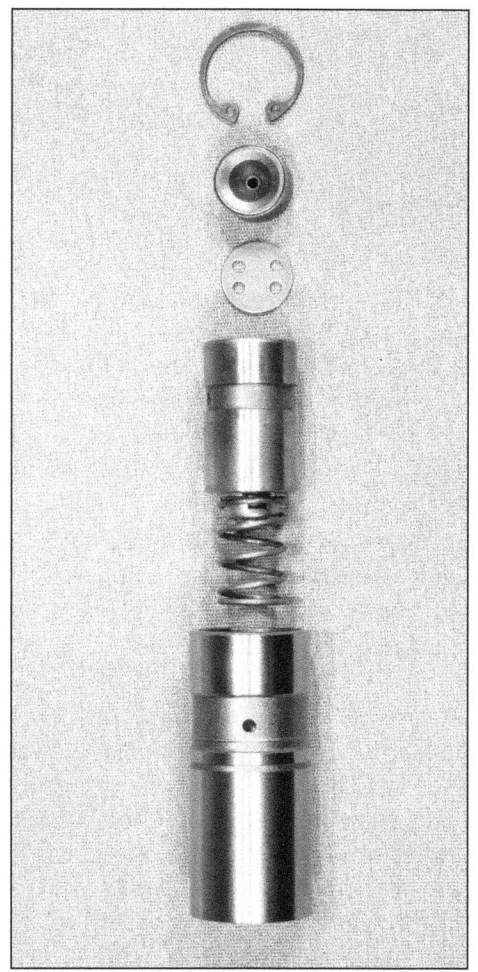

A hydraulic valve lifter is rather complex. It uses a combination of mechanical spring pressure and oil pressure to continually adjust the pushrod cup, taking up valve lash for quiet and consistent valvetrain operation. Some lifters have higher rates of adjustability than others. Expect reliable performance from the lifters sold by large camshaft manufacturers.

reacts against valvespring pressure. This bleeding is referred to as "leakage," and the rate at which a lifter leaks is controlled by using various degrees of lifter-body-to-plunger clearance during manufacturing. The leak rate of Pontiac's hydraulic lifters varied with the original application, and some, such as those for the Ram Air IV were specially designed to mimic the effects of a mechanical camshaft to maximize performance while maintaining a low-maintenance valvetrain.

Factory Cams

A camshaft opens and closes the valves at specific points during crankshaft rotation. Those values, along with a few others makeup the "timing" or "events" of a particular cam. The amount of time an intake or exhaust valve is off its seat is referred to as "duration" and it's expressed in degrees of crankshaft rotation. Since the crankshaft must rotate twice (or 720 degrees) to complete a single four-stroke cycle, the timing set gears are appropriately sized so that the camshaft operates at half the speed of the crankshaft.

Pontiac literature generally lists only the advertised duration for its camshafts. That can be recorded at whatever lift point Pontiac chose. Measuring duration at .050-inch lobe lift simplifies camshaft comparisons and gives a more accurate indication of the effects a particular camshaft has on total engine operation. The lifter fixture with dial indicator from Comp Cams is part of the camshaft degree kit I routinely use.

Pontiac's original rocker stud doesn't allow for manual valve lash adjustments. Its 7/16-inch shoulder tapers into 3/8-inch upper threads, and breakage can occur at the bottleneck with a high-lift and/or aggressive profile cam. To provide the durability required for its Ram Air IV, which featured valve lift greater than .500 inch, Pontiac used a 7/16-inch rocker stud, which necessitated manual lash adjustment using a crimp-style rocker nut.

Camshaft duration can be measured at most any point of lifter rise and fall. "Advertised" duration is typically the amount of crankshaft rotation recorded between .006 inch of lobe rise and .006 inch of lobe fall. Advertised duration can have different meaning for different manufacturers, however. Aftermarket manufacturers began providing duration specifications at .050-inch lobe lift so consumers could compare various cams and predict the performance effects. That particular value was selected because it's a reference point that easily measured and it's about the point that significant airflow begins.

Duration determines the operating range of a particular engine. A short-duration cam accentuates low RPM power. As engine speed increases, it has a limited amount of time to fill or exhaust its cylinders before the piston changes direction. Increasing duration allows better cylinder fill and evacuation at high RPM, which promotes maximum power production in the operating range the engine is heaviest loaded.

Common Pontiac Camshaft Specifications

Pontiac used a number of hydraulic flat-tappet camshafts in its V-8s. Most are identified using a single character stamping located on the face of the first or last journal depending upon the model year. The following chart contains the part number, identification stamp, and valve specifications of the Pontiac's more common camshafts including several of its performance grinds.

These camshafts are no longer available from General Motors, and a select few are available from various OE-replacement suppliers or certain aftermarket camshaft companies. If a stock-type camshaft best suits your project, most major camshaft manufacturing companies can produce a custom-spec unit for you.

Part Number	ID Stamp	Advertised Duration	.050-inch Duration	Gross Valve Lift at 1.5 Rockers	Lobe Separation	Intake Centerline
9777254	U	269/277	195/203	.375/.410	113.5	112.5
9779066	N	273/282	197/206	.410/.410	111.5	107.0
9779067	P	273/289	197/213	.410/.410	113.0	113.0
524886	6	283/293	207/218	.410/.410	114.0	112.0
9779068	S	288/302	212/225	.410/.410	116.0	113.0
9785744	H	301/313	224/236	.410/.410	115.5	113.0
9794041	T	308/320	230/240	.470/.470	113.5	112.0
493323	Y	301/313	224/236	.410/.410	115.5	113.0
549112	O	274/298	192/210	.365/.405	115.5	121.0
549431	(square)	274/298	192/210	.365/.405	112.0	116.0
10003402	Circled "A"	273/289	195/211	.395/.400	113.5	118.5

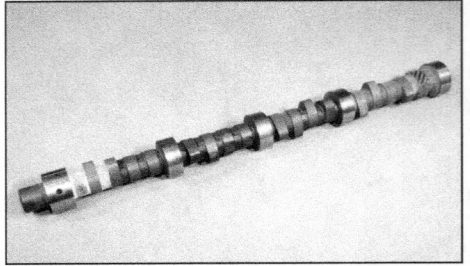

Pontiac used flat-tappet camshafts in its production V-8s. Constructed of a cast-iron core with hardened lobes, the factory cams were expertly chosen to maximize the performance of a particular application early on. When emissions concerns were of greater importance during the 1970s, performance was sometimes sacrificed to reduce emitted pollutants. The factory performance cams still provide excellent performance in modern builds.

Generally speaking, larger engines can tolerate more duration without negatively impacting low speed performance. The amount of camshaft duration a combination performs best with must match the size and intended operating range of the application. Other factors such as the position of the Intake Centerline (ICL) and Lobe Separation Angle (LSA) can be used to slightly improve idle quality or low speed street manners, or broaden the entire power band.

Pontiac's Engineering learned early on that cutting-edge camshaft design could put its production V-8s on the forefront of the performance market. It spent a great deal of time developing and testing various valve timing combinations to determine exactly what its engines performed best with. During the late 1950s and early 1960s, Pontiac's production camshafts were so well designed that oftentimes the factory-installed unit was capable of performance on par, or even slightly better than, many of the aftermarket units of the day.

Malcolm R. McKellar was one of the Division's top camshaft designers. In an interview before the retired Pontiac engineer's passing, he noted that the factory cams generally contained between 210 and 230 degrees of .050-inch intake duration, and several extra degrees of exhaust duration to compensate for the lesser exhaust port flow (when compared to the intake port) and the complete exhaust system. The lobe centers (or LSA) were generally spread several degrees apart to improve idle quality and allow an engine to be livable on questionable quality fuel. It's not uncommon to find a Pontiac camshaft with an ICL and LSA of 113 degrees or more.

Gross valve lift was limited to just over .400-inches for most applications. McKellar reasoned that Pontiac chose that particular amount because its cylinder heads generally contained rather large valves, and when combined with the small, high-velocity ports, high valve lift simply wasn't required to attain strong performance. The moderate valve lift allowed Pontiac to provide its customers with a good running vehicle that featured a low-maintenance valvetrain that was relatively free of durability issues over the life of the engine.

Pontiac used a number of hydraulic camshafts in its V-8 engines over the years. Even though it may seem as if two factory cams are virtually identical, each grind features some type of unique lobe

VALVETRAIN

With an adjustable valvetrain and hydraulic lifters, lash adjustments are accomplished by tightening the adjuster nut and spinning the pushrod with two fingers. Zero lash is the point where you just begin feeling pushrod drag. Depending upon the lifter manufacturer, recommended preload is generally between a half and one full turn past 0.

placement to complement its original application. The high-performance Pontiac cams hobbyists are most familiar with are the 068 and 041 grinds developed during the 1960s.

The 068 was developed for high-performance street use and contains 212/225 degrees of .050-inch duration and just over .400-inch valve lift. The 041 is a high-lift cam is most closely associated with the R/A IV, Pontiac top performance mill introduced in 1969. It contains 231/240 degrees of .050-inch duration and .470 inch of valve lift. Both are still available from OE-spec parts suppliers such as Melling and remain popular with performance enthusiasts looking for stock-type camshafts.

Factory Rocker Arms

Rocker arms transfer camshaft lobe lift into valve lift. The original Pontiac V-8 rocker arm, a stamped-steel unit pivoting on a fixed rocker stud, was developed and patented by Pontiac Engineer Clayton B. Leach. It was a very simple design that was so effective and cost efficient that many other automakers soon developed similar variants.

Camshaft Terminology

There are a number of terms used to describe camshaft valve events. Each has a specific purpose and a direct effect on performance and/or engine operation. The following definitions can help clarify those terms in a relatively basic manner.

Duration: The number of crankshaft degrees that the intake and exhaust valves are in motion between specific tappet rise and fall points. It commonly measured between .002- to .006-inch lifter rise (advertised), or at .020- or .050-inch lifter rise. The amount of duration is directly related to the operating range of a particular engine.

Lobe Lift: The measured distance between a camshaft's base lobe and its peak rise. It's generally stated in hundredths of an inch.

Gross Valve Lift: The theoretical total of lobe lift multiplied by rocker arm ratio. Additional valve lift allows a greater amount of peak airflow to enter the engine.

Intake Center Line (ICL): The crankshaft angle at which peak intake lobe lift occurs. It can be adjusted at the time of camshaft installation and dictates the crankshaft angle that the intake valve closes and the compression begins. Adjusting ICL can shift an engine's power range up or down.

Lobe Separation Angle (LSA): The angle between intake and exhaust centerlines. Stated in crankshaft degrees of rotation, it's a fixed value that cannot be altered. A narrow LSA tends to produce greater peak horsepower while a wider LSA tends to spread power over a greater range, improving average output.

Overlap: The period in which the intake and exhaust valves are both open. It occurs between the exhaust and intake strokes as the exhaust valve closes and the intake valve opens.

A camshaft degree kit, such as this one from Comp Cams, is an excellent tool that any hobbyist should learn to use. It's used to accurately measure and/or calculate all the critical valve events of a particular camshaft and to ensure that it's installed at the desired crankshaft angle. Verifying the installation point provides maximum possible performance, especially if slight adjustment is required.

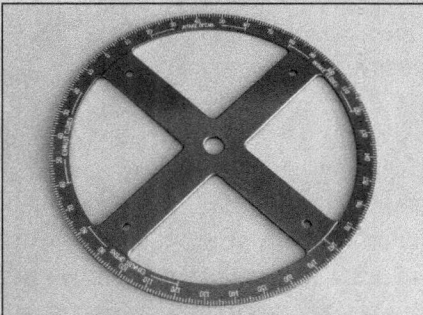

Most camshaft degree kits include an 8- to 9-inch degree wheel that's quite adequate. The larger diameter of a 16-inch degree wheel like this Proform (PN-67490, $190) example makes it easier to visually identify smaller differences when recording specifications. I much prefer using it when degreeing cams in my own engines.

CHAPTER 7

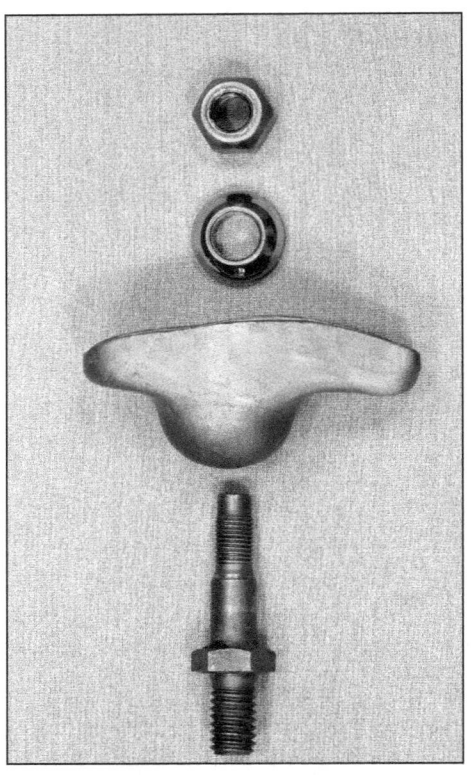

Developed by Pontiac Engineer Clayton Leach, the V-8 used a simple ball-stud rocker arm that was very efficient and cost effective. A rocker arm ratio of 1.5:1 was used on most production engines. Early Super Duty and Ram Air IV engines used an identical rocker arm but with a 1.65:1 ratio to increase valve lift.

Pontiac's rocker contained a 1.5:1 ratio, which allowed slightly more than .400-inch valve lift when combined with a lobe lift that's slightly greater than .270 inch. I have carefully measured several well-used originals and even an NOS unit and found the actual ratio closer to 1.48:1 on average. A 1.65:1 ratio rocker arm was developed for certain high-performance applications. Available through Pontiac's parts departments, it was a popular upgrade for many years after. Some aftermarket companies made low-quality knockoffs that offered a lesser ratio despite a claim of 1.65:1, but those I've measured from Pontiac were exactly 1.65:1.

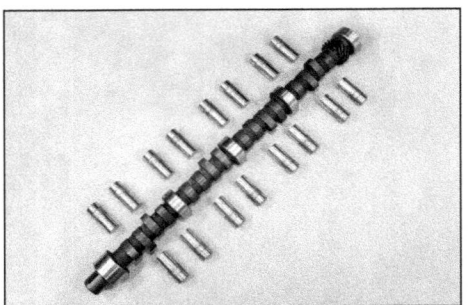

Flat-tappet camshafts remain very popular. They are produced and sold by a number of aftermarket camshaft manufacturing companies. Often overshadowed by modern roller cams, flat-tappets continue to offer an excellent dollar-per-horsepower value. They are capable of providing serious performance at a reasonable cost.

Aftermarket Hydraulic Cams

All Pontiac V-8s were equipped with flat-tappet camshafts. Hydraulic cams operate quietly and consistently, and the hydraulic action also compensates for production tolerances and normal wear. That action also acts as a cushion that absorbs a slight amount of lobe profile, which can slightly limit peak performance. Depending upon a particular lifter's leak rate, it might not leak enough at high RPM, and that can prevent the valves from fully seating, leading to a performance-inhibiting condition known as "valve float."

Despite those relatively minor shortcomings, aftermarket hydraulic flat-tappet camshafts remain quite popular with Pontiac hobbyists. Modern cam and lifter production techniques have essentially eliminated those concerns. A flat-tappet hydraulic cam is an excellent choice for any Pontiac engine regardless of the intended application, particularly if on a budget.

When combined with a high-quality lifter set such as that from Comp Cams, a modern flat-tappet is capable of producing a significant amount of high-RPM horsepower at a very reasonable cost. Generally speaking, a new cam and lifter set costs less than $250, and some companies such as Summit Racing even offer low-buck sets for those on a very limited budget. The low cost allows hobbyists to try several different units to determine which performs best with a particular engine.

Generally speaking, a typically-modified 467-ci Pontiac street engine might perform best with 230 to 235 degrees of .050-inch intake duration, but it's not unusual to find peak performance of a similarly-sized modern high-RPM race engine with 250 degrees of .050-inch duration or more, but you cannot expect the engine to operate well as speeds at less than about 3,000 rpm.

Aftermarket Solid Cams

Pontiac used manual-lash flat-tappet camshafts in its max-performance applications such as its SD-421. More commonly described as a "mechanical" or "solid" cam, it uses a lifter with a fixed pushrod cup that offers no hydraulic

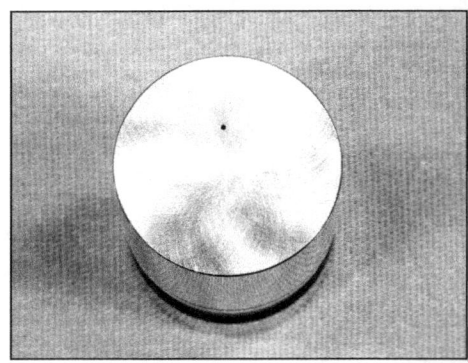

A solid-lifter cam offers no hydraulic cushion to minimize valve lash. The lifter has a fixed pushrod cup and simply passes oil to the pushrod for rocker arm lubrication. Crower's solid lifters are internally restricted, eliminating the need for block modifications. The Cool Face option includes a machined hole on the lifter face that directs pressurized oil between the lifter and lobe, positively lubricating the pair to reduce wear.

cushion. The intent is to allow the lifter to transfer the exact lobe profile to the valve, which can add several horsepower at high RPM while allowing the engine to operate more consistently at high speed.

The valves must be manually lashed and occasionally maintained. A feeler gauge between the rocker arm and valve tip is used to provide a specific amount of clearance gap that changes as the engine approaches normal operating temperature and the block and cylinder heads expand. An audible "tick" occurs as the gap is taken up during each revolution, which gives solid cams their distinctive sound, but it also consumes a certain amount of duration in the process. In fact, while it might first appear that an engine can tolerate more duration with a solid-lifter cam, the solid cam must oftentimes be 10 to 20 degrees larger at .050 inch to provide the engine with the same amount duration as a comparable hydraulic.

A hot lash setting around .025 inch is fairly typical, but the exact recommendation will vary with manufacturer and operating conditions. The various expansion rates of the materials used to construct blocks and cylinder heads will dictate how much lash should be added or subtracted to that recommendation to compensate for thermal growth when setting lash on a cold engine. Aluminum expands at a different rate than cast iron, so be sure to confer with the manufacturer to determine just how much cold lash is required for your particular combination.

Regular valve-lash maintenance was once required, and the perception continues to limit the popularity of solid camshafts for street applications. Modern rocker arm studs containing a flat flange and positive locking nuts have essentially eliminated the need for continual valve-lash adjustments.

While many camshaft manufacturers can provide you with a quality off-the-shelf or custom solid-lifter grind for Pontiac V-8, finding suitable Pontiac-specific solid lifters can be much more challenging. Hydraulic lifters require a large volume of pressurized oil for proper operation, and it also supplies the pushrods with oil to lubricate the rocker arms and valvesprings. A solid lifter, on the other hand, simply passes oil from the lifter bore to the pushrod.

Without limiting the supply of pressurized oil that reaches the pushrod, the oil pump can place more oil in the valve cover area than the engine can effectively return to the oil pan sump. That can cause oil to pool around the valveguides and enter the engine, or run the pan dry, effecting performance and/or reliability.

Pontiac used solid lifters that were internally regulated, but most aftermarket solid lifters in the past were modified hydraulic units that offered no internal regulation. Tapping the block to accept lifter bore restrictors was once commonplace, but Crower now offers Pontiac-specific lifters that are internally regulated to limit oil to the top end as well as offering its Cool Face option that positively lubricates the lifter and lobe. In my opinion, they are the only way to go if you're considering a solid-lifter cam.

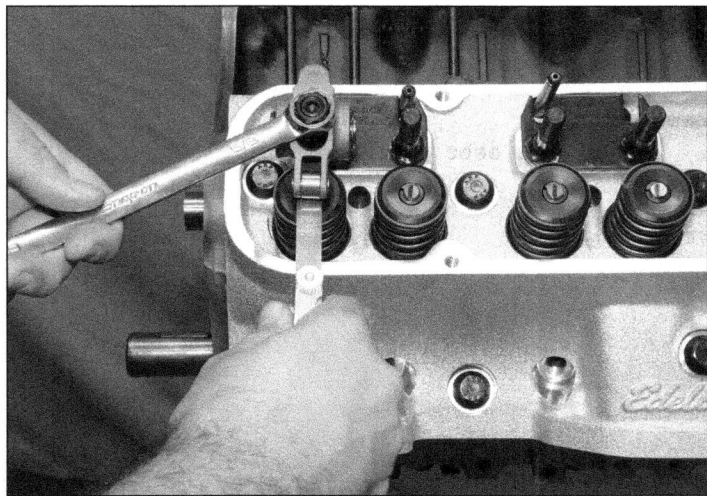

With a solid-lifter cam, valve lash must be set using a feeler gauge positioned between the rocker arm and valvestem tip. A certain amount of valve lash is required to allow for thermal expansion. The recommended amount varies with the components being used and the manufacturer. I recommend verifying valve lash seasonally. Minor adjustment is normal. Larger adjustments may indicate an underlying issue.

An adjustable valvetrain is required whenever the block and/or cylinder heads are machined moderately. Proform offers a special wrench (number-66781) that allows for quick and easy valvetrain adjustments when using positive locking rocker nuts. A long-handle 5/8-inch wrench tightens the rocker arm adjuster nut while a 7/32-inch hex-head wrench tightens the internal set screw independently. Once the proper lash adjustment is reached, simply tightening both simultaneously positively locks the rocker nut.

CHAPTER 7

Roller lifters use a larger roller wheel to reduce the sliding friction associated with a flat-tappet cam. The friction reduction allows quicker opening and closing rates for the camshaft lobe, which improves engine operation and performance. Hydraulic roller lifters continually adjust to minimize valvetrain lash and provide quiet operation. Reliability was once questionable, but modern high-quality lifters are much improved.

It's easy to see how roller cams improve performance. These two contain similar .050-inch duration and valve lift specifications, but the roller lobe (right) reaches peak valve lift much quicker and maintains it longer than the high-performance flat-tappet (left). A roller cam holds the valve open longer at higher lifts, allowing the engine to ingest more air to improve performance, but it also increases the amount of time the valve remains fully seated, which improves idle and drivability.

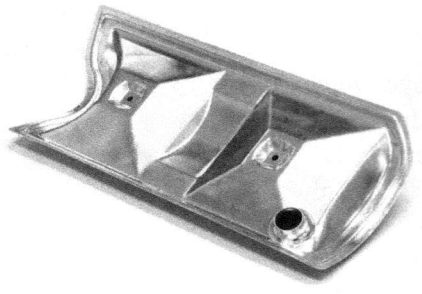

The valley pan is an integral part of a Pontiac's PCV system and oil vapors are continually passing through it. It's not uncommon to find heavy sludge deposits caked inside an original. Abrasives should never be used because small particles can hide in crevices and fall into the engine causing significant damage. The new Tomahawk valley pan from Pacific Performance Racing is an OE replacement that fits and functions like a stocker, but is modified internally to accommodate roller lifters. Raw sheet-metal valley pans are also available, but may not be compatible with the stock PCV system.

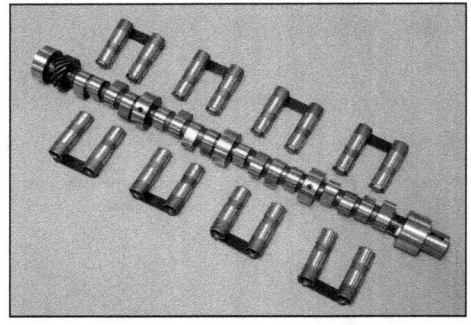

Hydraulic roller cams are very popular in modern performance builds. They increase performance while providing a quiet valvetrain that requires little maintenance. A cam and lifter set like this from Comp Cams costs several hundred dollars, but you get what you pay for. Comp's hydraulic roller lifers are among the best available today.

Aftermarket Roller Cams

Roller camshafts for Pontiacs have been around since the 1960s, but have become widely popular in recent years. That term can be somewhat misleading, however. The camshaft closely resembles a flat-tappet unit. It's the lifter body that's modified to accept a hardened roller wheel lending to the name.

The lifter's roller wheel reduces the friction associated with a flat-face lifter sliding across the lobe surface. In theory, that in itself can free up a few horsepower, but that's not always the case. The main advantage directly related to friction reduction is the ability to run a more aggressive lobe profile when compared to a similar flat-tappet. Opening the valve at a quicker rate translates into greater .050- and .200-inch duration without requiring additional advertised duration, and that can improve idle quality and low-speed street manners while maximizing peak performance.

Hydraulic roller cams have been around for several years. Combining those advantages mentioned above with hydraulic action produces a valvetrain that operates quietly and reliably for thousands of miles. Auto manufacturers began using hydraulic roller cams in production engines during in the 1980s and virtually all modern-production passenger vehicles use a roller cam.

Aftermarket camshaft companies began producing retrofit hydraulic roller cam kits for the Pontiac V-8 during the 1990s. The relatively heavy weight of the hydraulic roller lifter and the valve-spring pressure required to control it at high RPM caused the lifter to leak uncontrollably, which led to erratic operation. It took cam companies a few years to find a permanent solution. Many hobbyists developed a negative perception toward the reliability of hydraulic roller cams in that time, particular in high-performance applications. That's no longer the case, however.

Modern hydraulic roller lifters such as those available from Comp Cams, Crane Cams, or Lunati are much improved. You can expect such lifters to operate consistently and reliably up to 6,000 rpm, and

VALVETRAIN

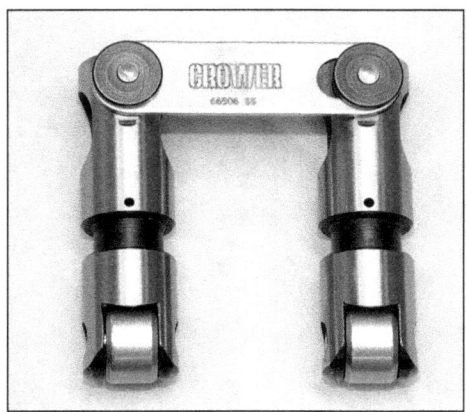

A solid-roller lifter acts like a flat-tappet lifter in that it has no hydraulic action to minimize valvetrain lash. The aggressive lobe profile associated with a solid roller cam maximizes performance. The solid-roller lifters for the Pontiac V-8 produced by Crower Cams are very popular with professional engine builders.

The quick lobe action of a solid-roller camshaft and the valvespring pressure required to control it are very hard on the roller wheel and its bearings. Comp Cams and Crower each offer solid-roller lifters for the Pontiac V-8 that includes a jet of oil that positively lubricates the bearing area. It significantly improves longevity in extreme applications.

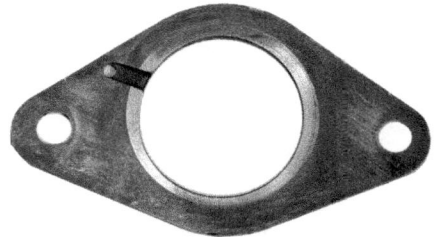

The camshaft retainer in a Pontiac V-8 also acts as a thrust plate, taking up forward thrust. After thousands of miles with a flat tappet, a slight amount of polishing where the camshaft rides is normal, and there should be no issue reusing an original in a performance rebuild. Wear like this should be considered excessive, however, and an original in excellent condition or a suitable reproduction should be used. The hardened steel core of a roller cam can accelerate retainer plate wear.

possibly even more. Beyond the initial cost, which can be as much as $800 or more for the cam and lifters, there really are no negatives to using a hydraulic roller cam in any street engine or a competition engine that operates at or near a maximum speed of 6,000 rpm.

Using my own 467-ci Pontiac as a test subject, I compared a few different high-performance hydraulic camshafts on the dyno to determine what performance benefits a roller might offer over a flat-tappet. Using camshafts with similar .050-inch duration specs and lobe placement, I found the roller improved idle quality and manifold vacuum, while noticeably improving midrange torque and slightly improving peak horsepower. I found that my engine operated better at all speeds and conditions.

I believe that the gains I saw were directly related to the aggressiveness of the roller lobe profile. When compared to the flat-tappet, the roller cam offered greater seat time (or less advertised duration) and contained nearly 20 degrees more .200-inch duration, which indicates a

Roller timing sets are commonly used during Pontiac V-8 rebuilds. High-quality units like this from Sealed Power (number-CTS-3112R) include durable sprockets and a high-quality roller chain with dual rows that resist excessive stretching. The crank sprocket generally includes more than one keyway, which can be used to slightly advance or retard the camshaft to improve performance. The set sells for about $150 and is quite suitable for mild to moderate performance builds.

high RPM, while the lesser advertised duration reduced valve overlap, which improved idle quality and vacuum.

Solid-roller cams provide the greatest potential for peak performance, but when using a camshaft with very aggressive lobe profile, it was not only hard on valve seats, it had the tendency to pound on the small roller bearings that separated the roller wheel and axle, which

CHAPTER 7

A few companies produce a roller timing set for the Pontiac V-8 that uses billet-steel gears to improve accuracy and durability. The crank sprocket usually contains a number of keyways for complete cam adjustably. Sims Performance Machining uses a Rollmaster timing set to produce its unique roller retainer set that's designed to reduce friction and improve reliability in extreme applications.

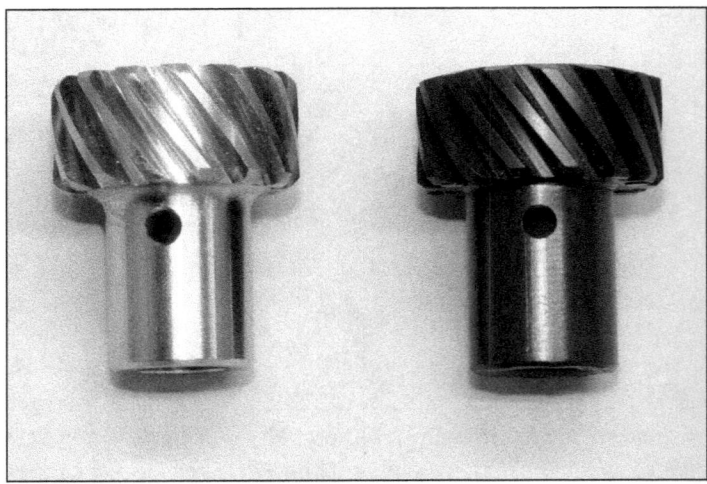

The distributor drive gear of a hardened steel roller cam isn't compatible with the iron driven gear typically found on original and aftermarket distributors. Failure results in sending metallic filings throughout your engine. Though a few cam companies offer a pressed-on iron drive gear as an extra cost option, it isn't very common. If you're using a roller cam in your Pontiac, a high-quality brass unit like that from MSD (left) or polymer driven gear like that from BOP Engineering (right) is your best option.

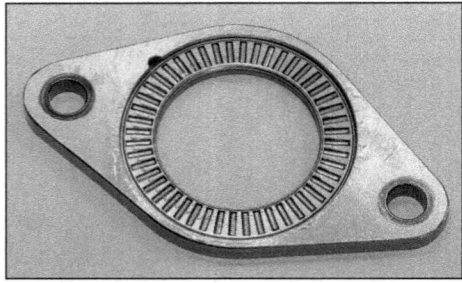

Sims Performance Machining developed a new camshaft retainer that's fitted with Torrington bearings. It's designed to reduce friction, improve durability, and limit camshaft thrust to maintain steady spark timing. It's an excellent concept that's available only from Sims. It sells for less than $275 and is compatible with any application, but particularly ideal for those that regularly operate under heavy load at high RPM.

eventually leads to failure, which can then cause significant collateral damage. Not enough valvespring pressure for the application can cause the lifter to bounce, resulting in a similar consequence.

A few companies producing solid-roller lifters for the Pontiac V-8 have found a solution that greatly improves reliability. Comp Cams, Crane Cams, and Crower Cams each offer Pontiac-specific lifters that direct a constant supply of pressurized oil to the roller wheel and axle assembly to positively lubricate the needle bearings. The lubrication cools and cushions bearings, significantly lengthening service life. Isky also produces solid-roller Pontiac V-8 lifters, and one is available with a low-friction roller that's free of needle bearings to maximize strength in the most severe applications.

Like Crower's flat-tappet solid lifters, most quality solid-roller lifters are internally regulated and do not require separate lifter bore restrictors to prevent excessive oil delivery to the top end. Since there are so many solid-lifter options, I suggest contacting the manufacturer to determine whether those you are considering are self-regulating and include high-pressure bearing oiling. Expect to spend $500 or more on a quality set of solid-roller lifters, and I am comfortable recommending any of those above.

Some hobbyists successfully run a mild-spec solid-roller camshaft on the street without issue. In those instances I recommend occasionally checking valve lash to ensure it hasn't deviated from the original setting, which may signal abnormal wear. If lash is within spec, it's unlikely that you'll encounter any significant issue as long as the valvespring pressure doesn't degrade. For a competition engine, I highly suggest inspecting the lifters seasonally. Most lifter manufacturers offer a rebuilding service where the lifters are thoroughly inspected and repaired accordingly at a very reasonable cost and in a timely manner.

Aftermarket Rocker Arms

An engine sees camshaft duration and lift at the valve through the rocker arm that pivots on a stud. Gross valve

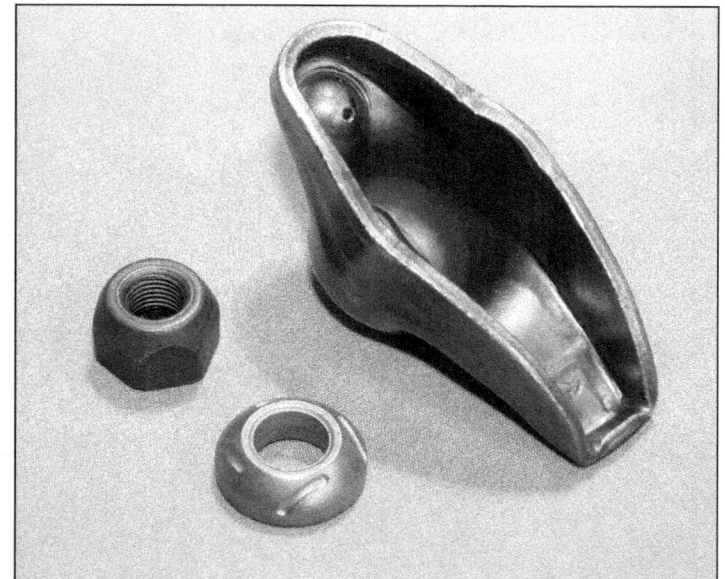

Certain OE manufacturers such as Federal-Mogul and Melling offer new stamped-steel rocker arms in a 1.5:1 ratio for the Pontiac V-8. The quality is generally good and they're practically identical to the originals. Melling offers this stock-replacement rocker arm with a ratio of 1.65:1 (PN MRK-532). I have used it in stock-type applications without issue, but recommend roller rockers where applicable.

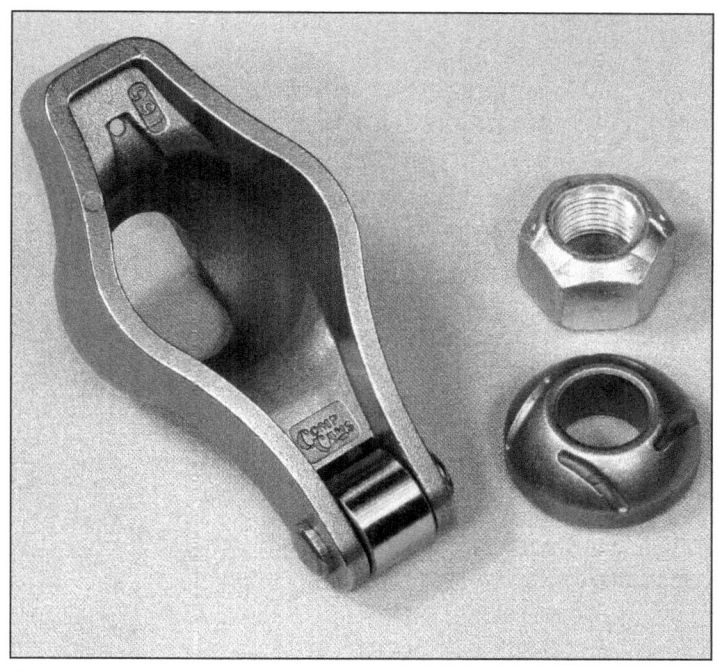

To improve the performance of the stamped-steel rocker, Comp Cams added a roller wheel to its tip to reduce friction and valvestem wear. Available in ratios of 1.52:1 and 1.65:1, the quality of Comp's roller-tip rockers is very good. I recommend them in any application where stamped-steel originals are being considered.

The pushrod cup's position in relation to the fulcrum determines the opening ratio of a particular rocker arm. The closer the pushrod is to the pivot point, the greater the ratio. It's much easier to understand exactly how it differs by setting two rocker arms with different ratios next to one another. A 1.5:1 ratio rocker is on the left while a 1.65:1 ratio rocker is on the right.

Full roller rocker arms combine a roller tip with a roller fulcrum to further reduce friction. Rocker arms constructed of billet or cast alloy are quite popular in mild to moderate performance builds. Low-buck roller rockers are generally of lesser quality and can fatigue and fail, but high-quality units like this from Comp Cams are affordable and reliable.

The high-pressure valvesprings required when using an aggressive roller camshaft can cause alloy rockers to deflect under load, leading to fatigue and failure. In those instances, a cast stainless-steel roller rocker arm like this one from Crower is an excellent choice that provides the durability required for high-performance use.

lift is calculated by simply multiplying camshaft lobe lift by the rocker-arm ratio. The closer the pushrod cup is to the pivot point (or fulcrum) the greater the resultant ratio. For production engines, Pontiac found it easier to achieve peak valve lift using moderate ratio rocker arms as opposed to greater lobe lift. Most production engines were fitted units stamped-steel units containing a 1.5:1 ratio.

When attempting to achieve a specific amount of gross valve lift, a high-ratio rocker arm is less stressful on the block when compared to lobe lift since the lifter travels less in its bore. Though it cannot alter the actual opening and closing points of a camshaft, a high-ratio rocker arm not only increases peak valve lift, it opens and closes the valves at a quicker rate, and that can make the camshaft appear a few degrees larger at the same time. A high-ratio rocker arm can improve performance in instances where an engine can tolerate slightly more duration or additional cylinder head airflow where available.

Original stamped-steel rocker arms, whether 1.5:1 or 1.65:1 ratio, are quite suitable for a stock rebuild, even in engines with mild performance enhancements. Comp Cams expanded on that rocker arm concept, adding a roller wheel in the tip of the stamped-steel body. With installation identical to an original, the roller tip reduces the sliding friction and side loading that can wear the valvestem and/or guide. PRW offers a unit similar to Comp's. Available in 1.52 and 1.65 ratios for Pontiac, such rockers sell for about $150 per set and are a good choice when replacing a set of stockers.

Most camshaft manufacturing companies offer full-roller rocker arms for the Pontiac V-8 in a wide array of ratios. Full-roller rockers generally consist of a cast-alloy or stainless-steel body that uses a roller trunion along with a roller tip to minimize friction. I have used cast-alloy rockers from Comp Cams and Crower

Shaft-mounted rocker arms are required for the most extreme applications. In contrast to individual studs, the rocker arms are grouped in pairs, and each pair shares a shaft. Each shaft is then fastened to a single rail, and the rail is fastened to the cylinder head. It eliminates deflection and provides maximum valvetrain stability.

Cams in various ratios with excellent results in high-performance street combinations, and would expect those from other reputable manufacturers such as Crane Cams and Harland Sharp to perform similarly. Expect to pay around $300 for a quality set.

When using high-pressure valvesprings, such as those required with an aggressive solid-roller camshaft, the cast-alloy roller rockers can fatigue over time. In such applications I feel it's worthwhile to use a rocker arm constructed of stainless-steel to maximize reliability and durability, and minimize deflection. Crower Cams produces a premium stainless-steel rocker for the Pontiac V-8 that's available in a variety of ratios and sells for about $500 per set. PRW offers a similar unit for those on a budget, and while quite suitable for myriad applications, I consider Crower's the best available.

A shaft-mounted rocker arm system is the best choice to maintain valvetrain geometry in extreme applications where valve lift approaches or exceeds .700 inch and where very high spring pressure is required. The rocker studs are replaced by a single rail that's securely fastened to the rocker arm flange of a cylinder head. Individual pairs of rocker arms pivoting on a common shaft are then fastened to the rail, essentially eliminating all flexing and deflection. Yella Terra produces a few off-the-shelf Pontiac offerings in 1.65:1 ratio for Edelbrock cylinder heads that sell for about $600. T&D Machine Products can custom make a set of shaft-mounted rockers in most any ratio possible with pricing that starts at $1,000.

Aftermarket Rocker Studs

With its 1.5:1 ratio rockers, Pontiac originally used a rocker arm stud that contains 7/16-14 base threads, and a 7/16-inch upper shoulder that tapers into 3/8-inch-diameter threads.

VALVETRAIN

T&D Machine produces a premium shaft-mounted rocker arm set for the Pontiac V-8. Each set is essentially custom made for the application and a wide array of ratios is available. T&D also has the ability to produce the intake rocker arm with its pushrod cup offset to accommodate wide-port cylinder heads.

ARP's 7/16-inch rocker stud is the best available today. Most original and aftermarket Pontiac cylinder heads accept ARP number-135-7101. It has a flat flange on top that's an ideal surface for the set screw of a positive locking rocker nut to tighten against. The stud design is so effective that valve lash rarely deviates from its setting even after extended periods of normal operation. Required adjustment may indicate a wear issue somewhere within the valvetrain that's unrelated to the stud itself.

Tightening the rocker nut to 20 ft-lbs locks it in place against the stud and automatically sets hydraulic lifter preload. Replacing the original "bottleneck" rocker studs with 7-16-inch units commonly associated with big-block Chevy engines is a common performance modification. A locknut is required and it allows manual adjustment of lifter preload, which can improve high-speed engine performance, but it also has additional benefit.

Pontiac used a similar manual-lash setup when it specified 1.65:1 ratio rockers to improve reliability and durability, or when solid lifters were originally used. Combining a high-ratio rocker arm with an original 3/8-inch rocker stud can shear off the stud top, causing significant collateral damage. Even though some aftermarket rocker arms are available with 3/8-inch locknuts, I strongly suggest using 7/16-inch rocker studs regardless of performance level. Those available from ARP are among the best available. Selling for less than $50 per set, it's cheap insurance against failure.

Valvetrain instability can lead to many operational issues. When exposed to very high valve lift and/or extreme valvespring pressure, rocker studs flex under load. A "stud girdle" is a machined fixture that attaches to a bank of rocker arms to improve rigidity and maintain geometry. The girdle clamps around a special set of positive locking nuts that are compatible with most aftermarket rocker arms. There are many aftermarket stud girdles to choose from and the cost is generally less than $100. Your Pontiac engine building specialist can help you decide if a girdle is necessary for your application. Aftermarket valve covers and/or a thick spacer may be required.

Aftermarket Pushrods

A pushrod is a length of metal tubing that transfers lobe lift from the lifter to the rocker arm, and its hollow center passes oil to lubricate the rocker arms and valvesprings. Because of its length, which is generally 8 inches or more in a Pontiac, it can flex slightly under heavy load, such as that created when using very high pressure valvesprings, and that can reduce valve lift. Many camshaft manufacturing companies produce high-quality pushrod sets that resist deflection, promoting maximum performance. I have had excellent results with the Hi-Tech pushrods from Comp Cams, but your Pontiac engine builder may recommend a similar offering.

The pushrods in any V-8 must be of the appropriate length to maintain optimal valvetrain geometry. The stock length of 9.135 inches very rarely applies when using popular aftermarket components, but particularly with a hydraulic roller camshaft. Comp Cams makes a very accurate and useful set of adjustable pushrod length checkers that allowing you to determine the exact

CHAPTER 7

length required. Once you achieve the proper geometry, you can then use that measurement to order new pushrods for your application.

Valvesprings and Valves

Valvesprings control valve motion. A valvespring compresses as the valve opens and must recoil controllably to keep the valve firmly on its seat without bouncing when closed. The amount of pressure a spring exerts when open and closed is determined by the number and diameter of its coil windings and the height at which it's installed. The required amount of spring pressure is mostly dependent upon the intended operating range of the engine, aggressiveness of the cam lobe profile, and overall weight of the entire valvetrain.

Pontiac originally used a dual cylindrical spring package and it was quite effective. It was retained by a large retainer and a pair of tapered locks. Aftermarket dual spring packages are available from a variety of manufacturers in a wide variety of pressure ratings. Quality sets from such manufacturers as Comp Cams,

Custom-length pushrods are almost always required when building a high-performance engine. Along with the camshaft, lifters, and rocker arms you intend to use, an adjustable pushrod length checker and low tension "test" springs installed onto an intake and exhaust valve is used to determine the appropriate length for a particular engine. Apply a dab of white grease on the valvestem tip and install the rocker arm and adjust the pushrod to a reasonable length. Rotating the engine several times by hand clearly establishes a contact pattern. The correct length leaves a tight roll pattern at the center of the valve tip. It may take several attempts to achieve the exact required length.

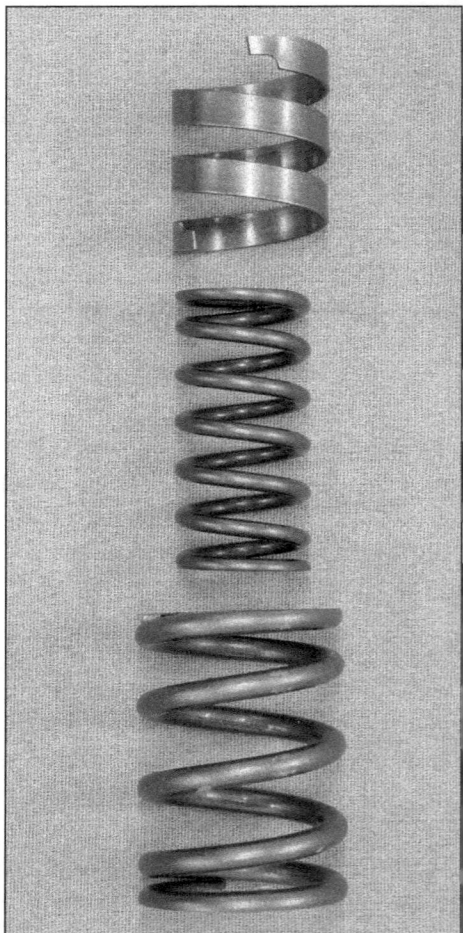

Pontiac originally used a dual-valvespring package in its production V-8 engines. Such sets are still an excellent choice and the aftermarket has many options containing various open and closed pressures and installation heights. Most aftermarket sets consist of an outer coil and inner coil that are separated by a damper, which absorbs harmonics.

What is Pushrod Bind?

Installing high-ratio rocker arms onto a Pontiac V-8 is a relatively simple process. Since the extra ratio is accomplished by relocating the pushrod cup inward slightly, it changes the pushrod's path angle, potentially causing it to contact the pushrod opening that's machined into the cylinder head. Nearly every cast-iron Pontiac head and many aftermarket aluminum castings require elongating the pushrod hole slightly when using a rocker arm with a ratio of 1.65:1 or greater.

A high-speed grinder and appropriate bit can be used to taper the upper half of the hole toward the rocker arm stud until sufficient clearance is gained. It's a task that should be performed with the cylinder heads removed from the engine, and is one your cylinder head assembler can easily perform for you during the porting process for a very reasonable cost.

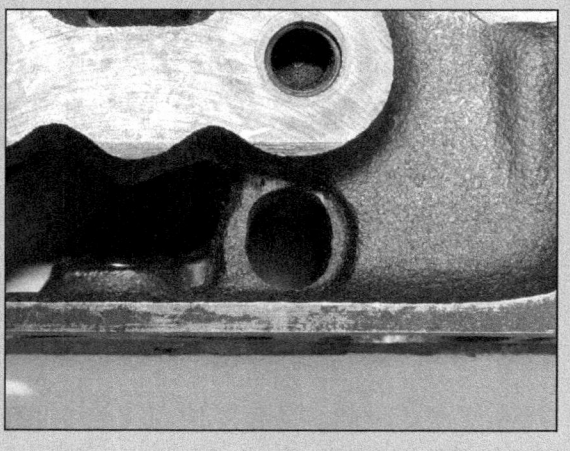

VALVETRAIN

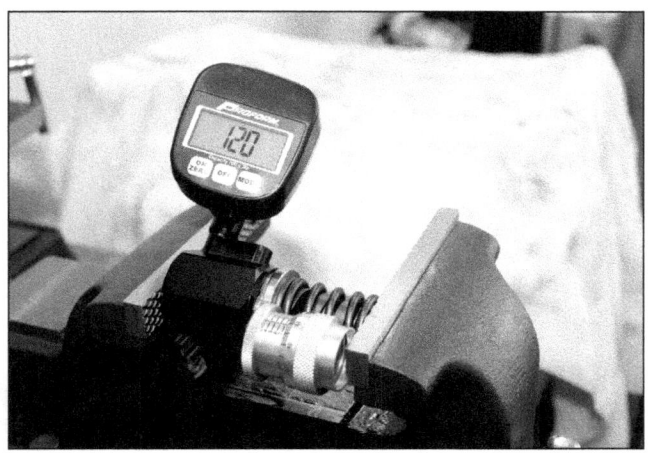

When combined with a Proform valvespring micrometer (PN 66902), it ensures accurate pressure readings at the appropriate open and closed measurements

Valvespring testing equipment is rather pricy, but is absolutely required to verify the valvespring pressure of new or used springs. Proform has recently introduced its digital mini valvespring tester (PN 66836) that's ideal for home use. It's an affordable option (under $200) that requires nothing more than a large-jaw vise.

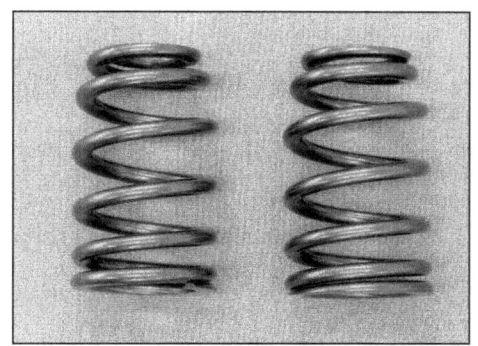

Conical (or beehive) valvesprings are growing increasingly popular. Tapering as it spirals toward the top, its shape better dissipates resonate harmonics to improve valvetrain stability while allowing a smaller retainer to reduce valvetrain mass. Its ovate wire provides sufficient spring pressure without the need for a second internal spring. While not required for every application, the use of conical valvesprings in modern Pontiac builds is widespread.

Crane Cams, or Crower Cams generally include an inner and outer spring, and internal damper. The damper is located between the springs and minimizes coil surge during operation.

Dual springs remain a very popular choice with Pontiac builders and aftermarket cylinder head manufacturers. They install easily with minimal modification and are available at a rather reasonable cost. A wide array of associated hardware is available for them and that includes stock-replacement retainers and locks, and high-strength and/or lightweight pieces constructed of tool steel or titanium. When purchasing ported or rebuilt cylinder heads from any specialist, be sure to discuss your valvetrain options to ensure the supplied valvesprings and ancillary components are compatible with your performance goal.

A conical valvespring, or "beehive" spring as it's often called, features a cylindrical shaped body that tapers as it reaches the top. The shape drastically increases spring rate and reduces the amount of moving mass within the spring, also allowing for the use of a smaller diameter retainer to further reduce weight. The uniquely shaped spring offers increased spring load while reducing the friction generated from added valvetrain mass

Valves constructed of stainless steel are commonly used in high-performance Pontiac rebuilds. Ferrea offers a variety of off-the-shelf units specifically designed for original and aftermarket Pontiac cylinder heads. It also offers valves constructed of its Super Alloy and titanium materials for max-performance applications in popular sizes. It produces custom valves in any dimension if required.

and component deflection. Your Pontiac engine building specialist can help you decide if such a spring is best for you.

Generally speaking, any time you purchase new cylinder heads, or send your existing heads to a porting professional, it will include a new set of valves

The stock Pontiac valve cover design doesn't leave much room for aftermarket valvetrain components. Taller-than-stock valve covers are available from a number of sources, but few are as nice as the aluminum units offered by Butler Performance. Available in a variety of colors, they're tall enough to accommodate most anything compatible with original or aftermarket Pontiac cylinder heads that use the stock layout. Valve cover spacers are available if additional height is required.

that are generally constructed of stainless steel or titanium, depending upon the application. If new valves are required, Ferrea, Manley, and SI are companies that offer top-quality valves for your Pontiac V-8 and may even be able to produce custom-spec units for you.

CHAPTER 8

INTAKE MANIFOLDS

The induction system of a Pontiac V-8 typically consists of a carburetor and intake manifold. Each is designed to maximize performance in a particular range. Larger race engines that operate at high speed require a significant volume of fuel and air, and that requires a unique intake manifold with multiple carburetors.

An intake manifold transfers the mixture of fuel and air from the carburetor to the cylinders. The plenum, or large voluminous area directly under the carburetor, gathers that combustible mixture before it's distributed to the cylinders through individual runners that connect to the intake ports.

It's tuned to provide maximum cylinder fill in a general RPM range and the design style and shape, and plenum and runners sizing strongly influences it. Engine displacement can also influence it. Larger engines tend to pull the power range downward, so a particular manifold that produces greatest average power from 3,000 to 6,000 rpm on a 400-ci, may work best from 2,500 and 5,500 on a 455-ci.

Intake manifolds can be classified into two distinct types, and it's directly related to plenum style. A dual-plane features a split plenum, with each half branching into four runners. The design allows half the engine to draw from half the carburetor at all times. A dual plane is most commonly associated with a stock manifold because it improves carburetor efficiency, which subsequently improves economy, emissions, and low-speed performance. It does, however, adversely affect peak performance at very high RPM.

A single-plane manifold contains a common plenum in which all eight intake runners draw. While it gives the engine greater voluminous area to pull from at high RPM, and can show strong performance gains in higher revving applications, a single-plane tends to degrade idle quality and throttle response in the process, and that impacts engines that are primarily street driven and operate at relatively low speeds. Which intake manifold type is best for a particular engine largely depends upon intended usage and operating range.

For its production engines Pontiac developed and used intake manifolds

INTAKE MANIFOLDS

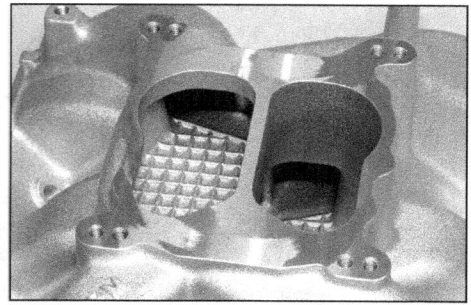

A dual-plane manifold splits the plenum in half. To maximize performance, the runners are staggered so the carburetor sees an equal amount of draw at all times, and that requires a top and bottom plane. The design improves air velocity through the manifold and carburetor, which improves engine efficiency at low to moderate speeds. A dual-plane is an excellent choice for a street engine.

A single-plane manifold contains a large, open plenum in which all eight runners draw from at all times. The design causes air velocity within the manifold to suffer at low engine speeds, affecting engine efficiency. Generally speaking, a single-plane manifold is best suited for engines that regularly operate at moderate to high speeds, and spend little time loaded at very low speed.

Pontiac used a dual-plane intake manifold constructed of cast iron and a 4-barrel carburetor on many of its passenger car engines in the 1950s and 1960s. The manifold provided excellent performance, but Pontiac used multiple carburetors and specific manifolds for its high-performance engines of the era. The original 4-barrel produced through 1967 is best left for restoration applications.

The use of multiple carburetors was banned by General Motors for 1967. Pontiac developed an all-new cast-iron intake manifold for its high-performance engines that model year. It was combined with the new Rochester Quadrajet carburetor to equal the performance of the Tri-Power it replaced. It performed excellently and remained in production through the late 1970s.

that accepted 2- or 4-barrel carburetors, and even those accepting multiple carburetors. Generally speaking, most were constructed of cast iron, but some special examples were cast in aluminum to shed weight and reduce intake charge temperature. There were even experimental intake manifolds produced for the early 1960s Super Duty that bolted together and contained a large common plenum that hobbyists referred to as the "bathtub."

Factory Cast-Iron 4-Barrel

When Pontiac introduced its V-8 in 1955, a 2-barrel carburetor was only available. An optional 4-barrel carburetor and corresponding dual-plane intake was introduced shortly after for performance minded enthusiasts. It continued on through 1966 for all 4-barrel applications and was originally topped by a square-bore-type Carter or Rochester carburetor depending upon the

The basic intake manifold casting was modified to accept emissions equipment over the years. An exhaust gas recirculation valve (EGR) was added to its passenger side in 1973. It remained in that position until traditional V-8 production ceased in 1978. Despite the negative perception, EGR didn't affect the performance potential of the castings.

A cast-aluminum intake manifold was developed to complement Pontiac's highest performance production engines of the late 1960s and early 1970s. Its runners were enlarged to accommodate the increased airflow capacity of the Ram Air IV and 455 H.O. engines it was used with. The aluminum construction improved heat dissipation. A separate cast-iron exhaust crossover (not shown) housed a divorced choke stove.

application and model year. Its coolant passages and mounting points were revised for 1965, making it compatible with popular Pontiac cylinder heads commonly used today.

Pontiac introduced a new 4-barrel manifold when multiple carburetion went away for 1967. It was a takeoff of the early 1960s Super Duty manifold developed for classes were only a single carburetor was allowed. It features long, smoothly contoured runners to maximize midrange torque and provide strong peak horsepower without compromising low-speed performance. Along with the new Rochester Quadrajet, the combination was used in all 1967 performance applications. The existing 4-barrel manifold and Carter carburetor remained for select standard performance applications, however.

The new manifold proved to be so well suited for myriad applications that it became Pontiac's only 4-barrel from 1968 forward, and the original design was discontinued. Over the years the casting saw the addition of an exhaust gas recirculation (EGR) system to quell certain emissions, and even physical deviations. It was discontinued after the 1978 model year, but several thousand were stockpiled for use in the optional 400 available 1979 Trans Am.

Factory High-Performance 4-Barrel

A specific cast-aluminum manifold was used with the 1969-1970 Ram Air IV and 1971–1972 455 H.O. An essential copy of the cast-iron manifold, the runners were raised about .125 inch to match the cylinder head intake ports and improve airflow. To isolate the manifold from excess heat, a separate cast-iron exhaust crossover connects the cylinder heads. Pricing can range from $400 to $1,500 depending upon the casting and if the original exhaust crossover is included. If considering one for your build, the 455 H.O. castings are typically most affordable.

Proper runner dimension and core shift can be common issues with the cast-aluminum manifolds, particularly the service-dated units sold though Pontiac's parts departments. Correcting takes little more than an hour with the grinder, and it's a process that's should be performed when using any intake manifold in a performance rebuild.

The 1973–1974 SD-455 cast-iron manifold is otherwise externally identical to a standard production casting that year, but its runners taller to improve airflow. It's easiest identifiable by the large "LS2" cast onto its coolant crossover and/or under-side. It rather rare and quite valuable, and an asking price of $2,000 or more isn't uncommon. Considering that it performs similarly to the cast-aluminum manifold, I suggest leaving these manifolds for purist restorations.

Factory Tri-Power

The Tri-Power manifold is quite likely Pontiac's most popular. A regular production option during the 1950s and 1960s, it performed consistently and reliably on the street. The center 2-barrel is used for idling, low-speed acceleration, and light-throttle cruising while the end carburetors open progressively as demand increases.

Most Tri-Power manifolds were constructed of cast iron. The 1965 and 1966 units are compatible with the original and aftermarket Pontiac cylinder heads commonly used today. The center 2-barrel carburetor is larger on the 1966 manifold, which increases its desirability and can improve performance, particularly with larger engines. Depending upon condition, a bare early-vintage manifold can

INTAKE MANIFOLDS

While a dual 4-barrel carburetor package was limited to Pontiac's competitions engines only, Tri-Power was a popular option that improved the output of its high-performance street engines during the late 1950s and 1960s. Fitted with three Rochester 2-barrel carburetors, the center unit is responsible for idling and low-speed driving. The end carburetors open progressively as engine load increases.

Any 1968–1972 Pontiac cast-iron intake manifold provides excellent performance in as-cast form. They can be purchased relatively cheaply and provide sufficient hood clearance with any Pontiac chassis. Castings from 1972 are most desirable, since its exhaust crossover is compatible with those on Pontiac cylinder heads produced during the late 1970s.

bring $100 or more. A 1965 manifold generally cost a few hundred dollars while the 1966 units can bring several hundred. Expect the 2-barrel carburetor trio and associated linkages and hardware to add several hundred dollars to that cost.

Factory Reproductions

A cast-aluminum reproduction of the original R/A IV and 455 H.O. manifolds is available from a number of sources. I haven't been impressed with the quality for performance use, the machining process leaves behind many sharp edges internally and plenum volume is reduced. A significant amount of grinding is required to achieve the flow capacity of an original. When considering the amount of time and money invested in buying and modifying a reproduction, it may negate the cost saving compared to buying an original, which might only require minor cleanup.

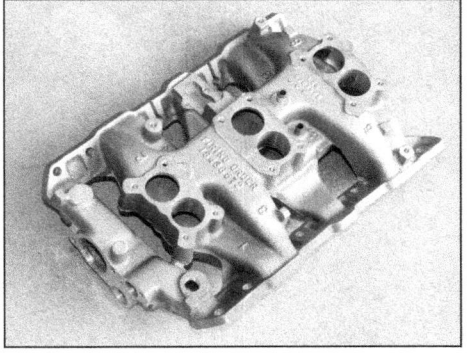

A reproduction of Pontiac's most desirable Tri-Power intake manifold has been on the aftermarket for a few years. It's a cast-aluminum copy of the 1966 unit that accepts all original equipment and is compatible with popular Pontiac cylinder heads. While it flows as well as an original in as-cast form, the aluminum construction allows for easy porting, which improves airflow capacity.

A cast-aluminum version of the 1966 Tri-Power manifold was produced in recent years. It's a near-exact copy of the original and accepts all corresponding factory accessories and hardware. The quality is generally very good and it's available from a number of sources. If you're considering a Tri-Power manifold for your build, I suggest using this piece. Ames Performance Engineering and Performance Years are two companies presently stocking new examples. Expect to spend around $400 for the bare manifold.

Selecting a Factory 4-Barrel

Any 1968–1972 cast-iron 4-barrel makes an excellent performance piece for Pontiacs operating up to 5,500 rpm or so. Just be sure the exhaust crossover passage is compatible with your particular cylinder heads. Prices generally range from 50 to $100.

The 1973–1974 manifold uses an EGR valve and contains a large passage beneath the exhaust crossover to feed it. Beyond being 5 pounds heavier, it performs as well as its predecessor and can often be found for less than $50.

The EGR system became an integral part of the 4-barrel manifold for 1975, and the secondary openings gained a pronounced "D" shaped appearance because of it. Despite the inhibiting appearance, it doesn't noticeably affect performance, but the negative perception makes such manifolds undesirable. They can oftentimes be purchased for $25 or less!

You probably noticed that I haven't mentioned the 1967 manifold, yet that's the year the new casting debuted. Its carburetor flange contains an exhaust passage intended to improve carburetor operation—a feature eliminated for 1968. The passage can be problematic if the appropriate carburetor gasket isn't used. I recommend plugging the holes and welding the corners to eliminate potential flange leaks. Since it performs identically to the 1968–1972 piece, I don't see spending any money or effort modifying a 1967 manifold unless it's required for a numbers-matching application.

Modifying Factory Manifolds

A few years back I compared the airflow capacity of several factory manifolds using a professional flow bench to determine which is best in stock form. Since a manifold and cylinder head work in conjunction, I measured the airflow effects an intake manifold had upon it cylinder head that otherwise flowed about 250 cfm. It could provide the flow potential of a particular manifold, which can be used to predict how it might affect performance.

Not surprisingly, 1969–1972 the high-performance cast-aluminum and 1973–1974 SD-455 cast-iron manifolds affected cylinder head airflow the least, flowing about 90 percent of the bare head. I found that all 1967–1978 standard production intake manifolds flowed similarly, regardless of flange shape or appearance. That included late-1970s casting with "D" shaped secondary bores. Total airflow of the standard performance castings averaged to more than 88 percent.

I then modified certain castings in an attempt to maximize airflow. Using various grinding bits and sanding rolls, I enlarged the intake runners to Ram Air gasket dimensions and maintained cross-section area toward the plenum to prevent any flow-inhibiting restrictions within the runner. I was able to improve the cast-aluminum manifold slightly to 92 percent, but the modification significantly improved the flow capacity of the standard iron casting, which totaled nearly 92 percent.

The result proved that after proper modification, any Pontiac 4-barrel manifold is a capable performer. If you'd rather forego the time and effort spent porting your own manifold, SD Performance has an excellent solution. It can CNC machine the runners and plenum

Smog-era manifolds are quite undesirable from a performance perspective, but don't discount those from the late 1970s if you're building on a budget or if class rules require it. An EGR valve was added in 1973 and the carburetor flange was modified in 1975, and the secondary bores gained a distinct "D" shape, but neither greatly affected flow capacity.

I measured the airflow capacity of many Pontiac intake manifolds on my flow bench. I used a cylinder head to establish a baseline figure and re-measured airflow with an intake manifold connected to it. In as-cast form, the high-performance aluminum manifold contained the greatest airflow capacity. The standard cast-iron manifold flowed slightly less.

INTAKE MANIFOLDS

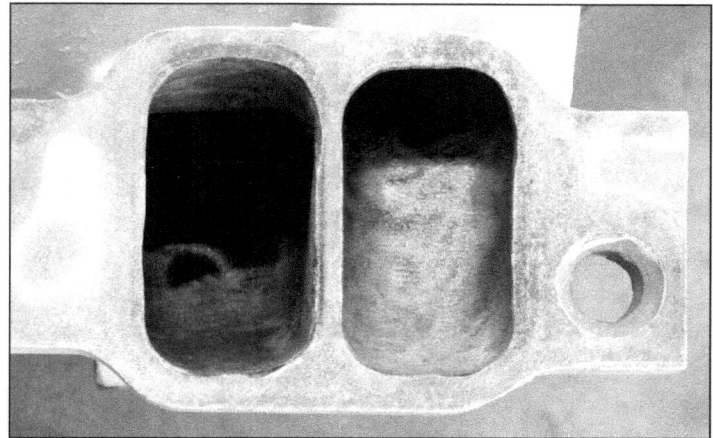

Using a cutting bit and high-speed grinder, I modified the runners of a few different Pontiac manifolds, enlarging them to full gasket dimensions in an attempt to improve flow capacity. A follow up session on the flow bench revealed definite improvements. The cast-iron unit showed the greatest gain, rivaling the airflow of the cast-aluminum unit.

If you're interested in improving the airflow capacity of your stock Pontiac intake manifold, but are uncomfortable performing the work yourself, SD Performance offers an excellent solution. It uses its CNC mill to enlarge the runners and modify the plenum of many factory cast-iron or aluminum manifolds. Optional hand finishing further enhances the effect. (Photo Courtesy SD Performance)

area of any original iron or aluminum casting, improving airflow. With pricing that starts at $175, the cost is quite reasonable, and there are a few options for even greater potential.

There's some debate to whether Tri-Power actually performed better than a 4-barrel in certain original applications. Pontiac actually installed a small baffle into some 4-barrel carburetors to limit airflow, giving Tri-Power a performance edge. I won't argue which is better, but am confident in saying that the Tri-Power manifold responds similarly to porting. If you wish to use Tri-Power on your engine and casting numbers aren't important, I strongly recommend the aluminum reproduction. It can make the porting process much easier than modifying an iron original.

Aftermarket Manifolds

There's a wide variety of cast-aluminum Pontiac V-8 intake manifolds available on the aftermarket. Generally speaking, they're designed to improve the output of modified Pontiacs, particularly at high engine speeds. While those accepting multiple carburetors were once rather common and certain examples are still available, single 4-barrel examples are most popular today.

A few aftermarket manifolds that are no longer in production are still popularly discussed on enthusiast web forums. Those include the Warrior, Edelbrock P4B and original Torker, and Holley Street Dominator. While still offering excellent performance, used examples can be found with some hunting. This book, however, only covers the popular aftermarket manifolds that are currently in production.

Edelbrock has been producing aftermarket Pontiac intake manifolds since the 1950s. Its stock-replacement Performer is a 50-state-legal dual plane that's compatible with most factory equipment and accepts both spread-bore and square-bore carburetors. Constructed of cast-aluminum to reduce weight, two specific variants are available—one with EGR for applications where it's required, and another without EGR. Pricing starts at less than $250.

In as-cast form, the Performer's runners are rather small. It's intended to maximize throttle response when used with smaller engines, but that can limit peak horsepower with larger engines producing more than about 400 hp. A significant airflow increase is possible by enlarging the runners to Ram Air gasket specifications, and while it's a modification I recommend when using the Performer in any high-performance engine, Edelbrock's Performer RPM may be a better choice in those instances.

CHAPTER 8

Separating the Coolant Crossover

A popular modification to any Pontiac intake manifold, whether original or aftermarket is separating the coolant crossover from the manifold body. It offers at few advantages, which includes easy removal without opening the cooling system, improved front-to-rear port alignment, and isolating the manifold body from coolant heat. The modification is quite simple and requires a few minutes with a sharp hacksaw blade and/or high-speed cutoff wheel to complete. A cutting bit and sanding rolls can be used to remove excess flashing and to smooth sharp edges if so desired.

With the exception of high-dollar castings or restoration applications, I routinely separate the coolant crossover on any intake manifold I use on my own engines.

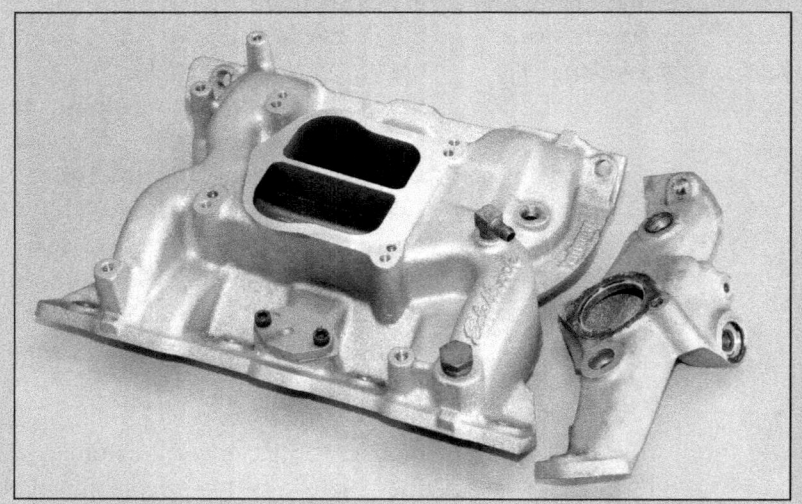

Separating the coolant crossover from the manifold body is a quick and easy modification that offers many advantages. It took about 15 minutes with a sharp hacksaw blade to cut through the flange and flashing of this Edelbrock Performer. If you're concerned with visual appearance, the excess flashing can easily be easily removed in a few minutes with grinding bits and sanding rolls. It's a modification I recommend for most performance builds.

The Performer RPM from Edelbrock is a high-performance dual-plane that provides excellent throttle response and low-speed street manners while allowing for excellent engine horsepower. It's much taller than a stock manifold and that can be an issue in certain applications, but it's the best aftermarket dual-plane ever produced for the Pontiac V-8.

The Performer RPM is a high-performance dual-plane that features a large plenum and very long, smoothly contoured runners. It provides an excellent balance of throttle response, low-speed street manners, and top end charge. The Performer RPM's carburetor mounting flange, which accepts both spread- and square-bore types is positioned about 1.25 inches higher than stock, and that can present hood clearance issues with certain Pontiac models. With a price starting at less than $250, I consider the Performer RPM among the best aftermarket intake choices available for street/strip applications today.

Edelbrock's original Torker was introduced in the mid 1970s as a high-performance single-plane compatible with the stock Quadrajet. Edelbrock replaced it with the redesigned Torker II in the late 1980s. The Torker II's carburetor flange is raised nearly .50 inch over stock. Its smoothly shaped runners connect to a moderately-sized plenum, which allows dual-plane-type throttle response and street manners at low speed, yet strong peak horsepower to nearly 6,000 rpm on larger engines. The Torker II only accepts a square-bore carburetor. It's an excellent manifold and sells for less than $250.

Pacific Performance Racing (PPR) introduced its single-plane Tomahawk a few years back. It's essentially a modern interpretation of Holley's Street Dominator, but with internal modifications to improve as-cast airflow. It's an excellent performer

INTAKE MANIFOLDS

Edelbrock's Torker II is a single-plane manifold that features a relatively small plenum with long, narrow runners that provide strong mixture velocity. The result is the ability to mimic dual-plane performance at lower speeds while maintaining excellent high-speed performance. The plenum volume is a bit small for larger engines. A carburetor spacer measuring at least 1 inch generally yields a performance improvement. (Photo Courtesy Edelbrock, LLC)

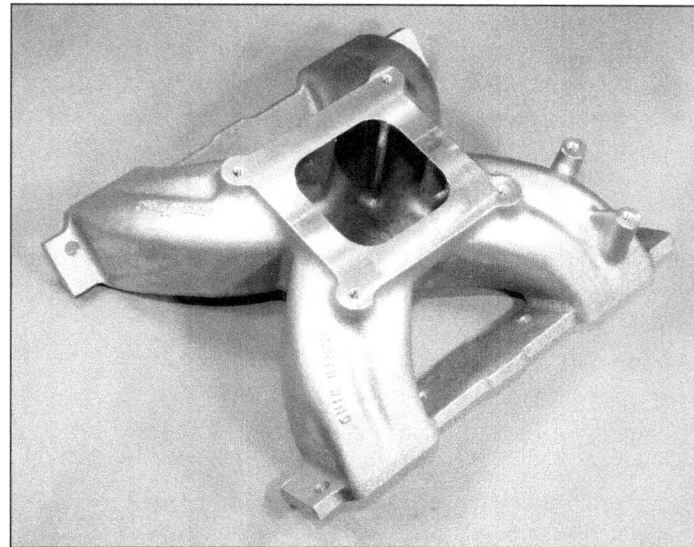

The Northwind manifold from KRE has a large plenum and runners, but its carburetor flange is only slightly taller than stock. It provides excellent performance without significant hood clearance issues, and plenty of room for porting. It's another excellent choice for competition engines.

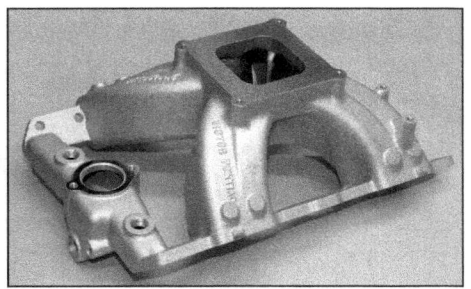

The Edelbrock Victor is a max-performance single-plane designed to promote maximum high-RPM horsepower. Its plenum and runners contain a significant amount of material that allows for port matching to a particular cylinder head. It's significantly taller than stock and requires some modification to the stock hoodline, but it's an excellent manifold for race engines with limited street use.

Certain cylinder heads have an intake port arrangement that requires a specific intake manifold design. Some manufacturers offer matching cast-aluminum intake manifolds, but others like this unit from KRE (for its Warp-6 heads), must be custom fabricated for the exact application. As long as hood clearance isn't a concern, the possibilities are limitless, including one or more carburetors.

for engines operating up to 6,000 rpm or slightly more. Its low-rise design and carburetor flange that accepts spread- and square-bore carburetors make it a popular option with second-generation Trans Am owners for its Shaker compatibility. Available from many vendors including PPR, pricing starts at about $200.

The Edelbrock Victor is a race-only manifold that can sometimes be used on the street. The high-rise single-plane was developed using input from professional Pontiac engine builders. It's often considered the best max-performance manifold available today. Its large plenum and runners contain plenty of material for custom porting so it can be tailored to the exact application. Two versions are available: one that accepts a typical square-bore carburetor and a one for use with a larger Dominator-type carburetor. Expect to spend at least $350 for either model.

The Northwind from KRE is another high-performance single-plane that's proven to be an excellent choice for max-performance applications. Its rather large plenum and runner sizes provides Victor-like performance, but its carburetor flange is lowered by more than an inch, which allows for greater adaptability to Pontiac models where hood clearance issues might otherwise exist. Available from many vendors including KRE, pricing starts at about $300.

The Hurricane and Crosswind from Professional Products are basically copies of Pontiac intake manifolds previously mentioned. I don't have any direct experience with their quality or performance level, however. There are a host of specialty manifolds specifically designed for use with certain aftermarket cylinder heads too. Those include the BOP, Tiger, Gutsram, CV-1, and Ram Air V units. While some may be compatible with traditional Pontiac cylinder heads, it's best to discuss compatibility with a Pontiac engine building specialist. Some companies can even produce a custom unit tailored to your application with single or multiple carburetors.

Carburetors

A carburetor atomizes fuel and mixes it with air flowing through it. It constantly varies the air/fuel mixture to produce maximum possible engine performance in all operating conditions. Carter and Rochester regularly designed OE-spec 2- and 4-barrel carburetors for Pontiac's production applications.

Rochester carburetors tend to be more popular. The model 2G 2-barrel was commonly used with the Tri-Power manifold. The Quadrajet 4-barrel introduced during the mid 1960s was first used by Pontiac for its V-8 in 1967. Few companies support the 2-barrel as well as The Carburetor

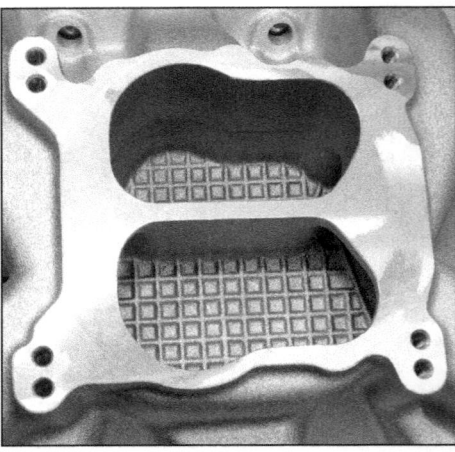

A spread-bore carburetor such as the Quadrajet has a different mounting bolt pattern than a square-bore type like the Holley. Some aftermarket intake manifolds have a carburetor flange with two-bolt patterns that accepts either carburetor type, increasing the number of carburetors you can use. Most high-performance aftermarket manifolds are designed to accept square-bore carburetors only, however. Check with the manifold manufacturer to determine if the unit you are considering is compatible with the carburetor you plan to use.

Pontiac commonly used the Carter 4-barrel during the 1960s. It was a decent performer but the lack of high-performance components and tech support limits its use today. While suitable for restoration applications, I don't recommend it for performance builds. There are other suitable options that perform as well or better.

The Rochester Quadrajet was introduced during the mid 1960s and it was first used by Pontiac on its V-8 in 1967. It's small primary barrels and larger secondary barrels provide an excellent balance of throttle response fuel economy, and strong full-throttle performance. It's quite popular with Pontiac hobbyists and those from the mid 1970s make excellent performance units when properly modified and can be purchased reasonably.

INTAKE MANIFOLDS

Holley has been producing carburetors for decades. Its 4150-style 4-barrel is likely the most popular high-performance carburetor in the aftermarket. The new Ultra HP–series 4150 4-barrel contains many new design features and is extremely adjustable, which allows owners and tuners to achieve maximum possible performance in all operating conditions.

Shop or Mike's Tri-Powers. I recommend either if needing quality replacement components for Tri-Power carburetors.

The Quadrajet uses small primary barrels to maximize efficiency and street manners during normal operation, and large secondary barrels to fulfill the airflow requirements for peak, full-throttle performance. The throttle bore size and spacing differences lend the term "spread bore." Pontiac used the Quadrajet on its V-8s ranging from 301 to 455 ci.

The Pontiac model 4MV used through 1974 generally offers a maximum airflow capacity of 750 cfm. There are, however, specialty units originally designed for the 1971 455 H.O. and 1973–1974 SD-455 that flow 800 cfm or more. The model M4MV debuted in 1975 and it contains an adjustable part-throttle (APT) feature that allows full mixture adjustment for optimal part-throttle operation. The M4MV used by Pontiac is capable of flowing more than 800 cfm with proper modification.

The Quadrajet is quite versatile, making it an excellent choice for high-performance rebuilds. Its main drawback is its rather small float bowl, which can be difficult to keep full without fuel delivery modifications. Cliff's High Performance supports the Quadrajet, offering quality stock-replacement and high-performance components. If you're considering a Quadrajet for your build, I recommend purchasing *How to Rebuild and Modify Rochester Quadrajet Carburetors* by Cliff Ruggles. It's an excellent resource with detailed rebuild and modification instructions for use up to 500 hp or more.

The Holley 4-barrel is likely the most popular aftermarket carburetor for engines producing 400 hp or more. Its dual float bowls contain the necessary volume to feed a high-horsepower engine for the entire quarter-mile, and its four equally sized barrels, which subsequently lends to its "square bore" description, also provide consistent fuel delivery at each corner. The 4150-series version flows as much as 950 cfm and is ideal for street/strip combinations. The 4500-series Dominator delivers as much as 1,250 cfm and is a popular choice for large-cube race engines.

The HP-series 4150 and 4500 carburetors remain very popular. Available directly from Holley or popular mail-order suppliers, expect to spend $600 to $900 for a 4150 unit, and $700 to $1,000 for a 4500 unit depending upon the model. Several carburetor specialists can customize a Holley by tailoring its fuel curve to match the requirements of your engine for extra cost. The Holley Ultra HP line offers the greatest adjustability to attain maximum possible performance on your own. The Ultra HP units generally sell for about $100 more than a comparable HP casting.

Carburetor Spacers

A potential performance increase is possible any time the carburetor is moved away from the plenum floor. A spacer allows the airflow exiting the carburetor to transition smoother into the runners

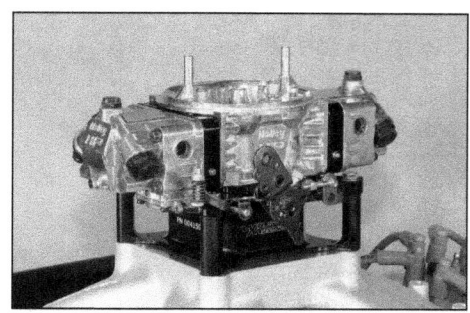

Carburetor spacers improve performance by lengthening the carburetor venturi or increasing plenum volume. Most Pontiacs show favorable results with a spacer as thin as 1/2 inch, but units measuring 1 to 2 inches generally provide maximum possible gain. Wilson Manifolds produces some of the best carburetor spacers on the market today.

while simultaneously adding plenum or carburetor venturi volume, depending upon the type. It can also insulate the carburetor from intake manifold heat.

Two types of carburetor spacers are available. An open-center spacer is best suited for a single-plane intake manifold as it adds plenum volume, giving the engine greater area to draw from, potentially adding several horsepower at higher RPM. Divided spacers are compatible with single- and dual-plane manifolds and generally contain four separate holes that effectively lengthen the carburetor bores. That can improve nozzle signal and carburetor efficiency, potentially adding horsepower and torque at low to moderate RPM, also pulling the power band downward slightly.

Spacer thickness can range from .500 inch to as much as 2 inches, depending upon availability, and performance generally improves as thickness increases. I suggest using the thickest spacer possible so long as hood clearance allows. Expect the greatest gains with a spacer measuring at least 1 inch thick, but I recommend testing performance with spacers of various thicknesses to determine what your

CHAPTER 8

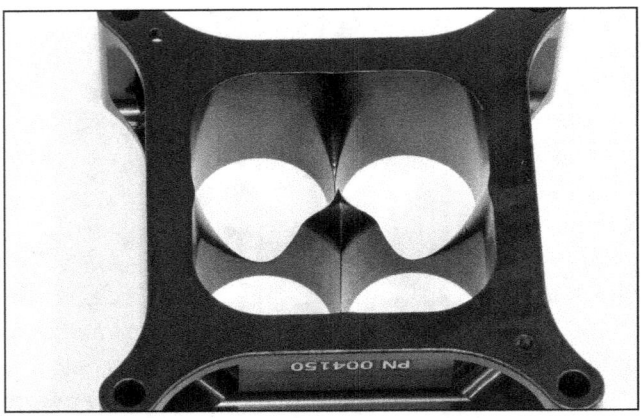

In addition to open-center, square-bore units, Wilson also offers carburetor spacers that are fully divided, containing four equally sized holes. Though traditional straight-bore spacers are available, the divider of Wilson's tapered-bore spacers smooth the transition from the carburetor into the plenum. It's an excellent design that's capable of providing a slight performance boost over the traditional design.

For those running a Quadrajet, Wilson offers an excellent spread-bore spacer that's fully divided and contains its unique tapered-bore design. It's a beautifully machined unit that noticeably improves engine output at all speeds. At 1 inch thick, it creates hood clearance issues with some Pontiac models, but it's certainly one to consider if you have extra room.

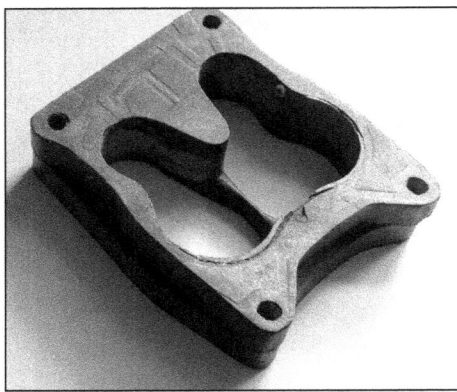

A carburetor spacer doesn't have to be pretty to be functional. Homemade units, such as this one belonging to noted Pontiac racer Jim Hand, was carved from two pieces of wood that have been glued together and shaped with fuel-resistant epoxy. The divided primary and open secondary simulate the effects of two different spacers in one package. It's worth 16 hp and 21 ft-lbs in a back-to-back dyno testing using a typical 455 producing 400 hp.

particular combination performs best with. It's not uncommon to find an increase of 5 to 15 hp or more depending upon the application.

Carburetor spacers are available from many sources. You can whittle your own from a piece of white pine wood or purchase readymade units constructed of laminated wood fiber, formed resin (phenolic), or aluminum for $20 to $100 from a mail-order supplier. In my opinion the best spacers available today are by Wilson Manifolds. Its tapered spacers further improve the transition from the carburetor into the plenum, enhancing performance. Wilson spacers sell for about $200, but I consider them money well spent!

Fuel Injection

Modern production vehicles have been using electronic fuel injection (EFI) since the 1980s. It's difficult to say if a fuel injection system provides additional full-throttle horsepower when compared to an appropriately sized and calibrated carburetor. The immediate fuel atomization and continual fuel mixture adjustments several times per second that EFI offers can improve engine efficiency at idle and low to moderate speeds on the street. The computer can also compensate for engine coolant and ambient air temperatures to provide smooth operation in all conditions.

While the Pontiac V-8 was never originally equipped with EFI, several aftermarket companies offer complete EFI kits that install relatively easily, allowing hobbyists to retrofit it. Some of the best EFI kits available today are produced by Edelbrock, F.A.S.T., Holley, and MSD. Each includes all the necessary components and wiring for complete installation, and the computer includes basic tuning for initial startup, and usually has the ability to adapt to the engine's fuel requirements as miles accumulate.

The Edelbrock EFI kit includes a Pontiac intake manifold designed to accept individual fuel injectors and high-pressure fuel rails. The F.A.S.T. system is similar, but doesn't include a manifold. (Edelbrock sells EFI-compatible manifolds to fit.) MSD's Atomic EFI is slightly different. It uses a throttle body with self-contained fuel injectors, and that makes it compatible with most aftermarket Pontiac intake manifolds that accept a square-bore carburetor. Expect to spend $2,500 or more for a complete EFI system.

Generally speaking, a carburetor (whether original or aftermarket) is usually compatible with the stock fuel system, and it can provide strong full-throttle performance so long as the mechanical fuel pump can keep the float bowl full at full throttle. The pressure and/or volume associated with the electric fuel pump typically required for EFI can be more than the rubber hoses and

INTAKE MANIFOLDS

Complete fuel-injection systems, such as the Edelbrock Pro-Flo, contain all the necessary components to convert your carbureted Pontiac to electronic fuel injection. The Edelbrock system includes the computer, all associated wiring, a large square-bore throttle body, and a modified intake manifold with runners designed to accept high-flow fuel injectors and a high-pressure fuel rail. A hand-held tuner is used to adjust the fuel mixture and spark timing to provide peak performance in all conditions. (Photo Courtesy Edelbrock, LLC)

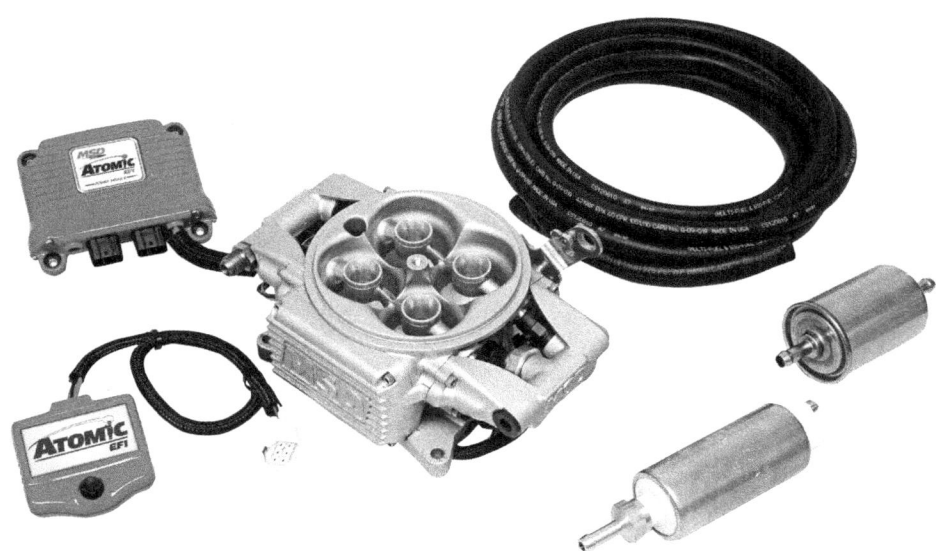

The Atomic EFI from MSD is an aftermarket fuel injection system that features a compact throttle body fitted with fuel injectors that simply and effectively replaces the square-bore carburetor. In addition to the self-contained throttle body, the kit includes a compact computer, various wiring harnesses, a pulse-width modulated fuel pump, and a sufficient supply of high-pressure fuel hose for easy and complete installation. (Photo Courtesy MSD Performance)

The throttle body included with the Atomic EFI system installs quickly and easily onto any aftermarket Pontiac intake manifold that accepts a square-bore carburetor and doesn't require a significant amount of wiring that can otherwise clutter up an engine compartment. When compared to the aftermarket carburetor it replaced, the MSD system improved the throttle response and performance of the 467-ci engine installed in owner Todd Ryden's 1965 Tempest Wagon. (Photo Courtesy MSD Performance)

steel lines of the stock fuel system can handle. An EFI system generally requires complete fuel system modifications, and that can add to the cost.

Forced Induction

Forced induction consists of mechanically compressing the air flowing through the engine. Forcing air into the cylinders packs more of it into a cylinder than what could otherwise be drawn in by a naturally aspirated engine. The result is a denser, tighter compacted air/fuel mixture that's more combustible and produces greater cylinder pressure upon ignition, yielding substantially greater performance.

Superchargers and turbochargers are popular forced-induction compressors. Boost is the amount of air pressurized by the compressor, and it's measured in pounds per square inch (psi). As the compressor moves a greater volume of air than an engine can consume, boost increases in the manifold. Too much boost can overstress gaskets and components, so a pressure relief (or blow-off) valve and/or waste gate is used to limit it during normal operating conditions.

Generally speaking, a supercharger uses long, twisted gears (also called twin screw types) or a centrifugal impeller to draw in and compress air. It's driven by a belt connected to the crankshaft snout and that allows the supercharger to delivers boost in a linear fashion that's consistent with engine speed. Driving the compressor does, however, consume a specific amount of engine power. Centrifugal superchargers are generally more popular because they're relatively small and efficient, and connect to the engine with brackets and tubes. ProCharger and Vortech are popular manufacturers.

A turbocharger looks and acts much like a centrifugal supercharger, but it's a bit more efficient as its impeller is driven by exhaust flow and not engine power. It operates at a higher temperature because of the exhaust stream, however, and that heat can transfer into the air that the engine consumes, negatively impacting performance. Turbochargers are generally cheaper than centrifugal superchargers, but the entire setup can be a bit more complex. Garrett and Precision Turbo are popular turbocharger manufacturers.

Superchargers and turbochargers come in a variety of sizes and flow capacities and which is best for you should be a decision you make with a professional engine builder. Any engine where forced induction will be used should be assembled accordingly, and a professional opinion can provide you with an idea of the components required and cost involved to reliably achieve your target power level.

Centrifugal superchargers are popular choices in forced induction builds. The compact design hangs from the engine and requires less underhood space. It's ideal for low-slung models where hood clearance issues may exist. This cutaway of a Vortech unit provides a clear look at the inner workings. Driven by a belt connected to the crankshaft, its impeller is geared to consistently provide a specific amount of boost at a given engine speed while consuming minimal engine power in the process. (Photo Courtesy Vortech)

A screw-type supercharger sits atop the engine and moves a much greater volume of air than a centrifugal supercharger, but it requires much more engine power to drive, and is a much larger package. It replaces the carburetor and intake manifold and requires some form of fuel injection. Screw-type superchargers are what professional Pontiac racers use most often for their competition engines.

A turbocharger operates much like a centrifugal supercharger, but its impeller is instead driven by exhaust gas. It is more efficient because it doesn't consume engine power during operation, but can transfer heat into the air it's compressing for engine ingestion. Turbochargers come in a variety of sizes and flow capacities, and a professional opinion can provide what's best for your budget and particular application. (Photo Courtesy Don Keefe)

CHAPTER 9

EXHAUST

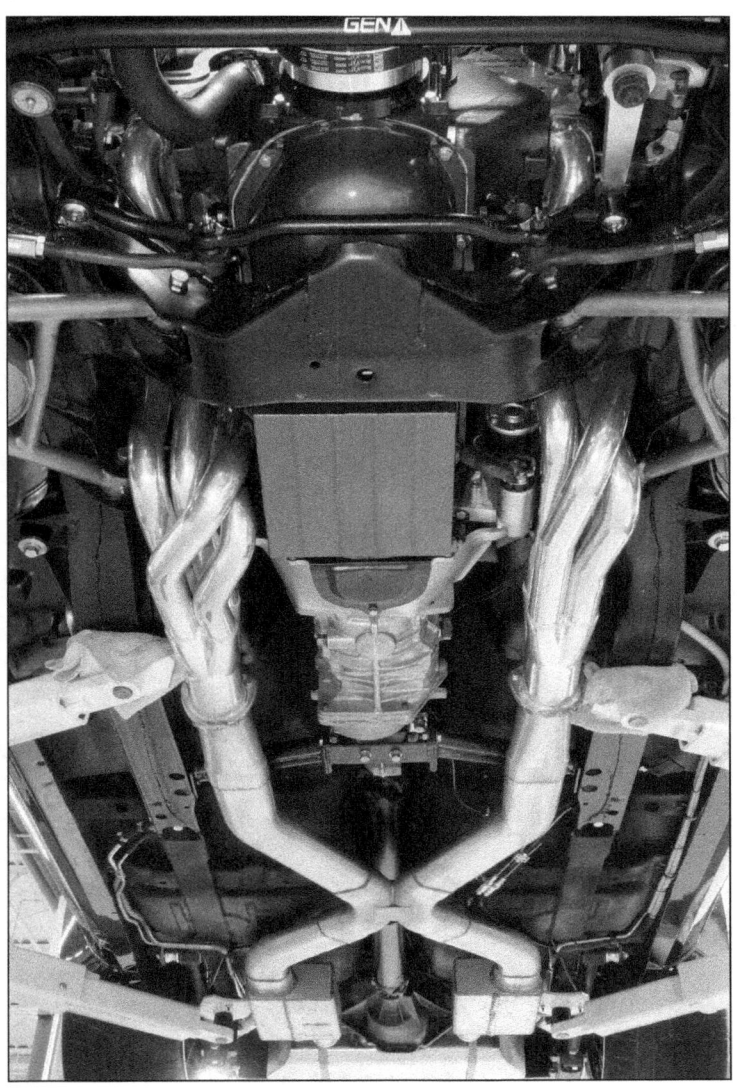

To attain maximum possible performance from any engine, a vehicle's exhaust system should provide the least amount of restriction. It should include tubular headers, large-diameter exhaust tubing, free-flow mufflers, and usually some type of crossover pipe. A properly installed system should tuck up tightly against the floorpan to provide the greatest amount of ground clearance.

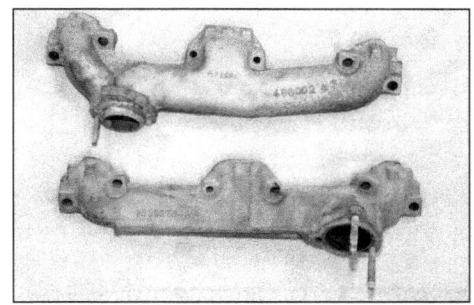

A cast-iron exhaust manifold that simply collects exhaust gas and routes it to the head pipe was used on most standard-performance Pontiac engines. These log-type manifolds, as they're often referred to, offer no exhaust tuning or pulse scavenging to improve performance. Only available in a D-port configuration, I don't recommend them for any application beyond a stock rebuild.

Factory exhaust systems were designed to fit within the space constraints of a particular chassis while maintaining a specific noise limit at a reasonable cost. While that might not detract from the performance of a typical passenger car application, additional performance at every speed can be attained by replacing the factory components with high-performance aftermarket pieces, including tubular exhaust headers,

CHAPTER 9

larger-diameter piping, and free-flowing mufflers. And that tends to increase noise.

A low-restriction, dual-exhaust system to allow greatest possible performance was made available in 1956 and commonly used through 1974, at which point a single exhaust catalyst was added, ending the possibility of using true dual exhausts.

High-Performance Exhaust Manifolds

As engine displacement increased throughout the 1950s and 1960s, Pontiac began experimenting with high-flow exhaust manifolds that featured long, individual runners that merged into a large collector area to improve horsepower, particularly at high RPM.

Basically a cast header constructed of iron, it featured a bolt-on collector that had a separate flange that could be uncapped to bypass the remaining exhaust system to achieve maximum performance in competition settings. Available through dealership parts departments, Pontiac's high-flow cast headers became part of its factory-installed Super Duty package introduced in 1962.

The cast-iron headers were very heavy. A cast-aluminum version was developed to reduce overall vehicle weight for those regularly competing in drag races. The cast-aluminum header was significantly lighter and performed suitably, but aluminum alloy technology and the casting and heat-treating processes were in their infancy. Durability issues were common in vehicles that were operated for extended periods. Claims that molten aluminum would

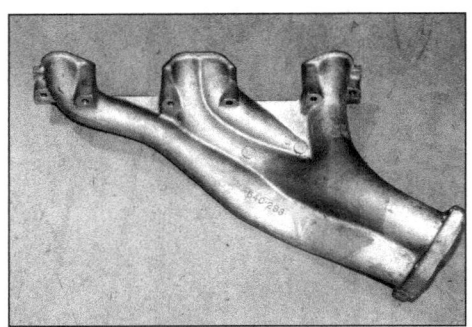

Pontiac developed a "tuned" header constructed of cast-iron for its early Super Duty engines and it was an excellent piece. It performed as well as any tubular header of the era. Its only drawback was weight. A cast-aluminum version was produced for weight-conscious racers. RARE manufacturers cast-iron and aluminum reproductions.

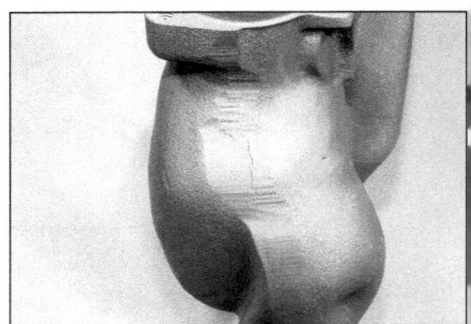

Exposure to a countless number of heating and cooling cycles throughout an engine's lifetime makes original cast-iron exhaust manifolds prone to cracking. Unless required for a numbers-matching application, I don't recommend using original high-performance manifolds in any build. Excellent reproductions are available.

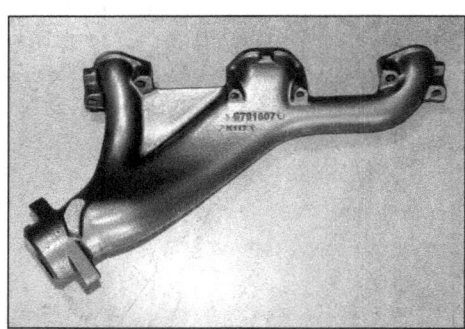

An abbreviated version of the cast Super Duty header was developed for Pontiac's high-performance street applications. The "long branch" manifold fits many full-size Pontiac models from the 1960s including the Grand Prix, as well as first-generation Firebirds. RARE offers excellent reproductions with 2.5-inch collectors. A Super Long Branch version is available with a 3-inch collector.

A streamlined exhaust manifold was developed for the 1967 400 H.O. available in the GTO that model year. It was used with many other Ram Air engines over the years, as well as the 1973–1974 Super Duty 455 (shown). The "Ram Air" manifolds, as they're often called, resemble "shorty" headers and fit tightly against the block for maximum clearance. They're a popular choice for many A-Body applications as well as second-generation Firebirds.

EXHAUST

literally drip from the exhaust system may be a bit exaggerated, but I have seen the dividing wall between the center ports eroded away.

A slimmer version of Pontiac cast header was developed for high-performance street applications during the early 1960s. These "long-branch" manifolds, as they're commonly called, feature long individual runners like the original cast header, but its collector was an integral part of the casting, which was flanged to mate to exhaust piping. The design was compatible with all full-size Pontiacs produced during the 1960s and the new Firebird introduced in 1967.

The long-branch manifold wasn't compatible with the intermediate A-Body platform, so Pontiac developed a streamlined exhaust manifold that debuted in 1967 on the 400 H.O. It became synonymous with Pontiac's A-Body-spec Ram Air engines beginning in 1968, and many commonly refer to it as an H.O. or Ram Air manifold. It was used through 1972 in certain A- and G-Body applications and many 1970–1974 Firebird models as well.

The factory high-performance exhaust manifolds are an excellent choice for Pontiacs that are primarily street driven. Along with quiet, leak-free operation, you're likely to find at least one type that fits your particular chassis. The original castings are quite old, however, and it's sometimes difficult to find crack-free examples. While original castings can sometimes be purchased reasonably, reproductions are available.

Reproduction Factory Manifolds

A few companies reproduce Pontiac's most popular high-performance exhaust manifolds. In my opinion, those from Ram Air Restoration Enterprises (RARE) are the best.

RARE offers exact reproductions of Pontiac's cast header originally used in early Super Duty applications in iron or aluminum. It also offers near-exact cast-iron reproductions of Pontiac's long-branch and Ram Air manifolds with D-port and round-port configurations.

RARE also offers a Super Long Branch manifold which features a 3-inch collector for easy connection to a high-performance exhaust system. Its Ram Air manifold is available with an oversized collector area measuring nearly 2.5 inches for added performance. RARE's cast-iron manifold generally sells for $400 to $600 depending upon the casting. I highly recommend them any time original high-performance exhaust manifolds are considered.

Tubular Headers

Exhaust headers produced from round exhaust tubing offers the least amount of exhaust restriction, providing the greatest potential for power output.

Primary tube and collector diameters can affect horsepower and torque. Larger-diameter tubes can improve airflow at very high RPM, but it can sacrifice charge velocity, which can negatively affect torque output, particularly at low speed. The best header for a large-cube Pontiac that's primarily street driven typically contains a primary tube diameter

At least two companies offer near-exact reproductions of Pontiac's most famous exhaust manifolds. They continue to be a popular choice for many high-performance builds where a stock appearance is desired. They are available from most restoration parts supply houses specializing in Pontiacs for a reasonable cost. Ceramic coating (top) adds extra cost.

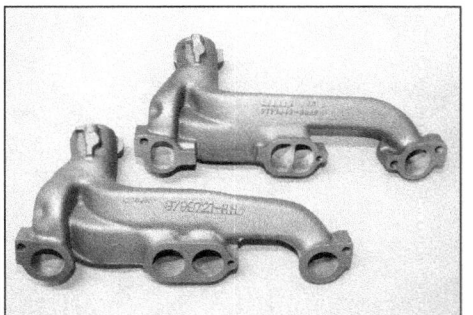

With the exception of the cast Super Duty header, Pontiac's high-performance exhaust manifolds were originally available in D-port (top) and round-port (bottom) configurations for use with corresponding cylinder heads. Modern reproductions are available in both styles and they are not interchangeable. If you plan to use reproduction manifolds with aftermarket cylinder heads, be sure verify with the manufacturer which configuration is correct.

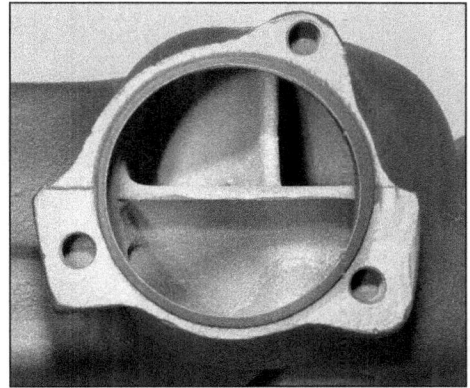

RARE offers its reproduction Ram Air exhaust manifolds with a larger-than-stock outlet size to improve airflow. It's a very popular option that adds several horsepower to most engines and is recommended if you plan to use original-style manifolds in your build. Be sure that the head pipes are appropriately sized for the manifold outlets.

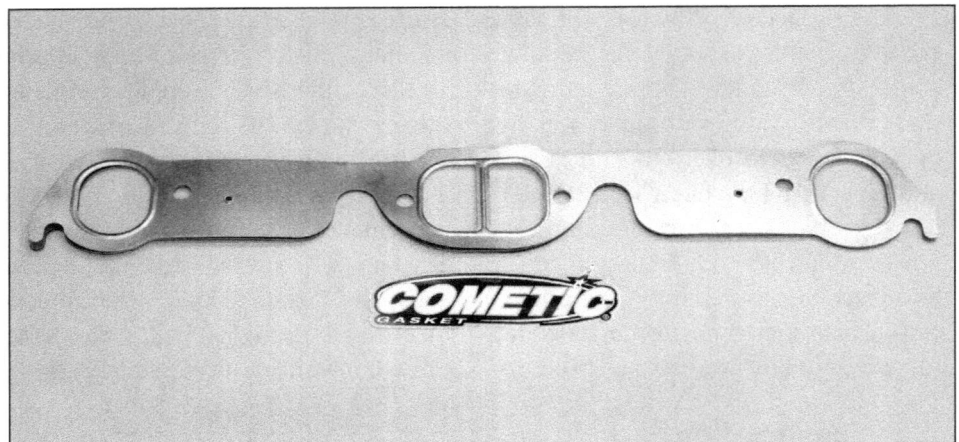

Gasket leaks are quite common with headers. A thick cylinder-head flange is ideal because it's very rigid and doesn't distort when tightened. That allows an even gasket seal, which can otherwise blow out when exposed to large volume of high-pressure exhaust, particularly when considering the varying rates of thermal expansion as the components reach normal operation temperature. Cometic produces an excellent multi-layer steel exhaust gasket that's intended to prevent leaks in performance engine.

Depending upon the chassis, some headers do not allow sufficient room for the stock oil filter housing and/or stock-size filter. It's very close on this example, and accessing the filter with the engine installed may be challenging. A small-diameter filter is one option, while a remotely located filter may be a better choice if space constraints do not allow the stock setup.

There's little need for a complete exhaust system on a dedicated race car. Instead, the header tubes dump out into normal atmosphere to provide the least amount of exhaust restriction. Such headers are typically custom made to fit the exact application, and some snaking may be required to fit the particular chassis.

EXHAUST

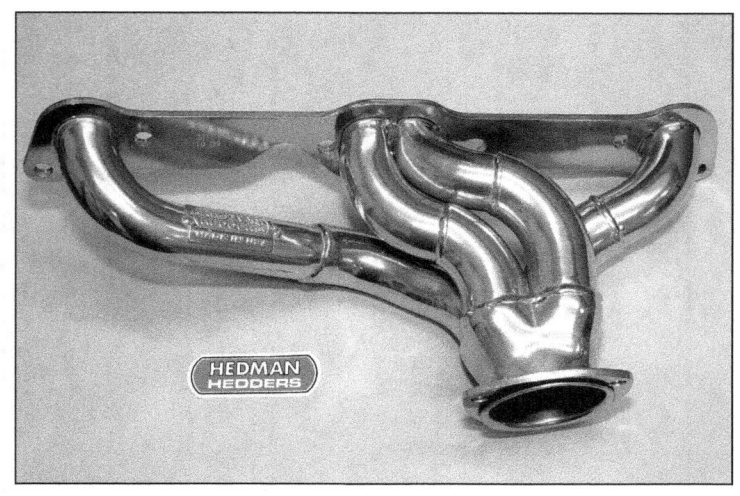

Hedman produces a shorty-style header for GTO and Firebird applications. On a dyno it performs similarly to a RARE Ram Air manifold with oversized outlet, but it weights substantially less. That benefits weight-conscious racers that are working with limited chassis space. Like other Hedman products, its shorty header is well made and has a very thick cylinder head flange to prevent gasket leaks.

Four-tube headers offer the greatest performance potential as they offer the least amount of restriction to exhaust flow. They are, however, rather large and can be difficult to squeeze into a tight chassis. Installation may require raising one side of the engine and/or inserting them from below. I recommend purchasing headers from a popular manufacturer. The quality and consistency is generally very good.

between 1.75 and 2 inches, and a collector diameter of 3 inches.

The size of your engine, cylinder head airflow, and the RPM range you plan to operate it in will dictate which size is best for you. There are a number of companies producing high-quality Pontiac headers today, including Doug's, Hedman, Hooker, and Mad Dog. Headers from such companies generally sell for $400 or more, but fit well, have a thick cylinder head flange to protect against gasket leaks, and include a flanged collector to accommodate the installation of a complete exhaust system. Your Pontiac vendor can suggest a set that's best for your Pontiac model and particular engine application.

There are a few different header types available, including shorty, three-tube, four-tube, and tri-y. Each has a specific purpose and price point.

Shorty Style

Shorty headers fit and perform similarly to Pontiac's Ram Air exhaust manifold. Hedman offers a D-port version in its Hedder line for GTO and Firebird models that features 1.625-inch-diameter tubing that steps to 1.75 inches to improve low-RPM torque, and 2.5- and 3-inch collectors. Selling for about $300 per set with a black paint finish, they're very well made and are an excellent alternative to Ram Air manifolds when attempting to save weight. Hedman offers a high-temperature ceramic coating that adds about $200 to the cost.

Three-Tube Style

Three-tube headers look similar to full-length four-tube headers. These contain separate end tubes, but the center tube is large enough to collect exhaust from both exhaust ports at the center of a Pontiac cylinder head. The center tube sacrifices performance, but it's easier and cheaper to produce, giving hobbyists an affordable tubular header that's usually easier to install because its three tubes take up less space.

Several companies offer three-tube headers. While they may be adequate for a low-performance rebuild, I don't recommend them for high-performance use.

Four-Tube Style

The best-quality headers have four primary tubes, one for each exhaust port on a particular cylinder head. In a four-tube design, the primary tubes snake through the chassis and merge into a single col-

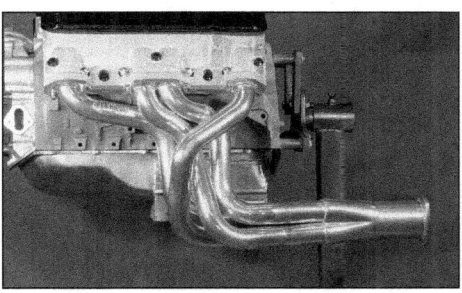

Doug's Headers is a very popular manufacturer. It offers full-length four-tube headers for many popular Pontiac applications. They're among the best fitting units available today and can be purchased reasonably from your favorite Pontiac vendor or large mail-order supplier.

lector, and it generally promotes greatest peak horsepower. Depending upon the header manufacturer, primary tube diameter can measure from 1.625 inches to as much as 2.125 inches, and collector diameter can range from 3 to 3.5 inches.

Tri-Y Style

A tri-y header has four primary tubes, which are gathered into pairs. The pairs then merge into a collector, which leads into a secondary tube, and the two

Four-tube headers consume much under hood space. That places the tubes very close to important accessories such as the starter. The heat source causes a large, stock starter to overheat and fail, and the tube routing makes it impossible to remove or replace without disconnecting the header from the engine and/or raising the engine to gain clearance. An aftermarket high-torque starter like that from IMI is a compact design that dissipates heat much quicker than the stock unit. I recommend such a unit anytime full-length four-tube headers are used.

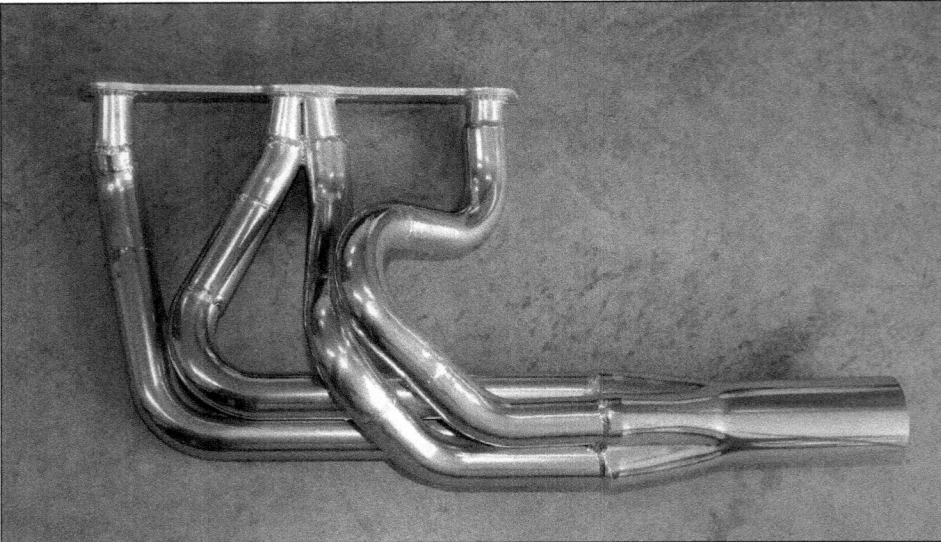

Ceramic coating not only provides an attractive finish that's very durable, it also acts as a thermal barrier that prevents exhaust heat from dissipating too quickly, slowing the charge as it exits. In addition to silver ceramic coating for tube headers, natural cast-iron (gray) is a popular color for cast exhaust manifolds.

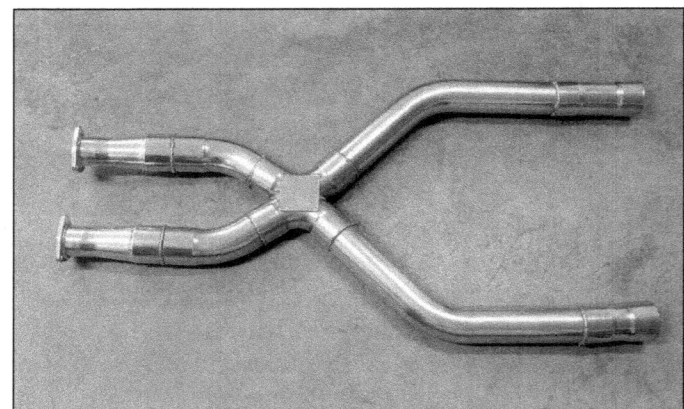

An X-type crossover creates a split junction that merges the head pipes, allowing the exhaust pulses from one to draw off the other, improving exhaust scavenging, and engine performance. It also gives the exhaust charge two paths in which to exit, effectively increasing the capacity of the entire exhaust system.

secondary tubes then merge into one large collector. The design acts much like traditional four-tube headers at high RPM, but it improves cylinder scavenging, which tends to improve mid-range power numbers. They are an excellent compromise that works very well on high-performance street engines.

H-O Racing developed an excellent tri-y header that fits many popular Pontiac models. Available with D-port and round-port exhaust configurations, it was quite popular during the 1970s, 1980s, and 1990s. It has since been discontinued, but used sets in excellent condition occasionally appear on Pontiac-specific classified lists or internet auction sites.

Ceramic Coating

Ceramic coating exhaust manifolds and tubular headers is a popular process. In addition to leaving behind a durable, high-quality finish that doesn't discolor when exposed to exhaust heat, it creates a thermal barrier that limits dissipation of exhaust charge heat through the pipe as the charge makes its way toward the muffler and tailpipe. That allows the charge to maintain a greater degree of velocity and that can improve performance slightly. It also limits the amount of heat that radiates from the manifold or header, reducing underhood temperature.

Many companies offer high-quality ceramic coating. Expect to spend more than $200 for coated cast-iron exhaust manifolds and even more for full-length tube headers. If you do not find someone to do this locally, some shops provide mail-order service. I am familiar with Trail Performance Coating in Omaha, Nebraska, and can recommend its services. The quality of its work is excellent and pricing is quite affordable.

X- and H-Type Crossovers

The addition of an exhaust crossover in a dual-exhaust system can improve the

EXHAUST

A large-cube engine moves a significant volume of air, and the diameter of the exhaust tubing must be great enough to support it. It's not uncommon to find a system with tubing 3 inches or more, which is required to maintain peak performance. Round tubing is popular but significantly reduces ground clearance on low-slung models.

scavenging that occurs as exiting exhaust pulsations draw the exhaust charge from adjacent cylinders. The scavenging lessens the amount of work each piston must do when forcing the exhaust gas from the cylinder and into the exhaust manifold or header. That can yield a definite performance improvement. There are two types of crossovers commonly found on performance vehicles: H and X. Determining which your Pontiac performs best with, if any at all, may require some trial and error.

An H-pipe is a balance tube that connects two head pipes. It can be constructed of a length of tube that's the same diameter as the head pipes or slightly smaller. It allows a certain degree of scavenging, but its greatest benefit is balancing the pressure pulsations within each head pipe. That can smoothen the sound emitted at the tailpipes, and can improve performance slightly.

An X-pipe merges the two head pipes at a junction. It improves exhaust scavenging but also gives the exhaust charge two paths in which to exit. I have found that an X-pipe offers at least some advantage in most engines. I've seen a slight improvement of midrange torque and peak horsepower on my own Pontiacs with an X-type crossover when compared to not having one. The X-pipe also produces a smoother, higher pitched exhaust note that sounds finely tuned.

Mufflers

Factory mufflers were designed to attenuate a specific exhaust tone and maintain a manageable noise level in and out of the car. Most high-performance mufflers are designed to produce a more aggressive sound while maximizing exhaust system flow. The aftermarket mufflers from Dynomax, Flowmaster, Magnaflow, Pypes, and Spintech are among the most popular with Pontiac hobbyists.

Many muffler manufacturers have sound clips on their websites. I suggest visiting those sites or attending local shows or cruises to determine the sound you're looking for.

Complete Exhaust Systems

Several muffler companies offer a complete exhaust system for many popular Pontiac models. It generally includes high-flow mufflers and large-diameter exhaust tubing that's aluminum coated or constructed of stainless steel and pre-bent for the application. Along with the muffler manufacturing companies

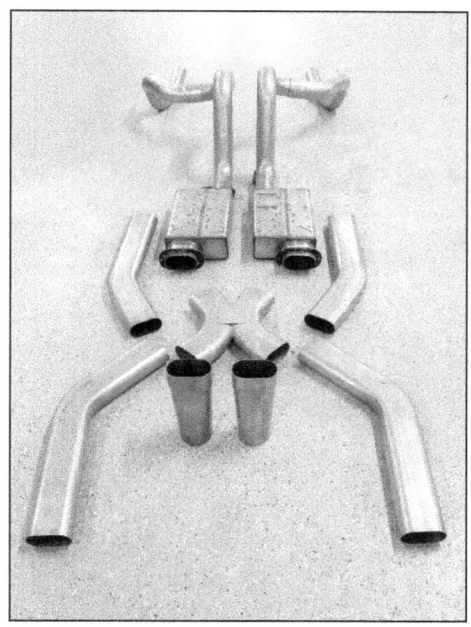

Round exhaust tubing measuring 3 inches or more may be impractical for some applications. RARE offers a complete exhaust system that features oval tubing entering the muffler and round tailpipes to maximize available space. The oval tubing provides the same amount of total area as a round pipe, but it doesn't sacrifice valuable ground clearance.

mentioned above, Ram Air Restorations Enterprises (RARE) also offers complete Pontiac exhaust systems.

When considering a complete exhaust system for your Pontiac I recommend one with the largest-diameter tubing. A diameter of at least 2.5 inches is mandatory, but I suggest 3-inch tubing if possible. It gives the exhaust charge a greater area to expand and dissipate its heat before it reaches the mufflers. Even larger tubing is available, but its overall diameter can present clearance issues, particularly with a low-slung chassis.

Some companies produce oval tubing that provides the required area without the clearance issues. Tailpipe diameter isn't as critical since the charge has lost most of its heat and velocity by the time it exits the mufflers.

CHAPTER 10

IGNITION

A vehicle's ignition system is responsible for igniting the mixture within each cylinder at the appropriate time. It generally includes a coil that saturates and stores energy and a distributor that routes spark to each cylinder at the appropriate point in the crankshaft rotation. Pontiac used a few different systems that all performed satisfactorily for production engines. While they still perform suitably, there are aftermarket offerings that may be better suited for modern high-performance engines.

Spark Lead

Maximum possible engine torque occurs when cylinder pressure peaks at a crankshaft angle of 10 to 12 degrees after top dead center (ATDC). That allows the pressure to expend the greatest portion of its energy over the entire length of the crankshaft stroke. Since it takes a fixed amount of time for the flame front to spread across the combustion chamber, ignition must be initiated at a crankshaft angle before top dead center (BTDC) to achieve maximum cylinder pressure ATDC.

Initial timing is the base spark setting without any vacuum advance. It's generally a setting between 10 and 20

The ignition system is responsible for providing the spark that ignites the combustible air/fuel mixture within the cylinders. The distributor routes electrical voltage to the spark plugs at the appropriate crankshaft angle for optimal performance. Pontiac used a few different distributors in its production engines and a wide array is available on the aftermarket.

degrees depending upon the application. As piston speed increases, spark timing must occur earlier in the rotation cycle and a mechanical assembly within the distributor increases spark timing. Total maximum spark lead in a naturally aspirated Pontiac V-8 is generally between 30 and 40 degrees, and that can vary with combustion chamber type, compression ratio, and fuel octane. The mechanical or "centrifugal" advance within a distributor can be adjusted by swapping around springs and the weight and center cam assemblies.

When an engine is under very light load, its carburetor throttle plates are

IGNITION

As engine speed increases, spark must occur earlier to provide optimal performance. A distributor contains an internal mechanism that uses centrifugal force to advance spark timing. It's a very simple setup that's easily adjusted by replacing the small springs, weights, and/or center cam, depending upon the distributor type.

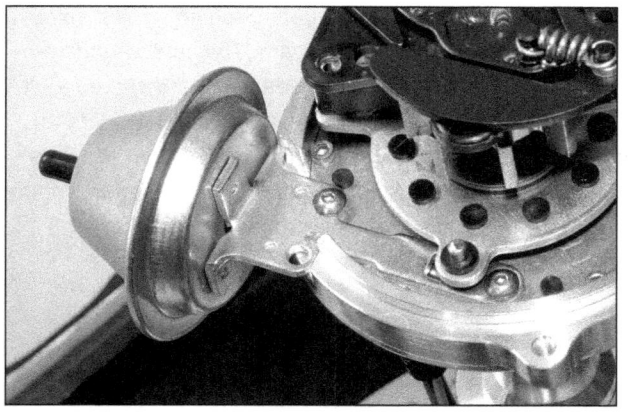

A vacuum advance canister senses engine load and adds timing in certain operating conditions to improve engine efficiency. That reduces the engine's operating temperature and increases fuel economy. Many hobbyists are intimidated by vacuum advance, but it's simple to use and its effects are quite noticeable. I strongly believe every street engine can benefit from at least a slight amount of vacuum advance.

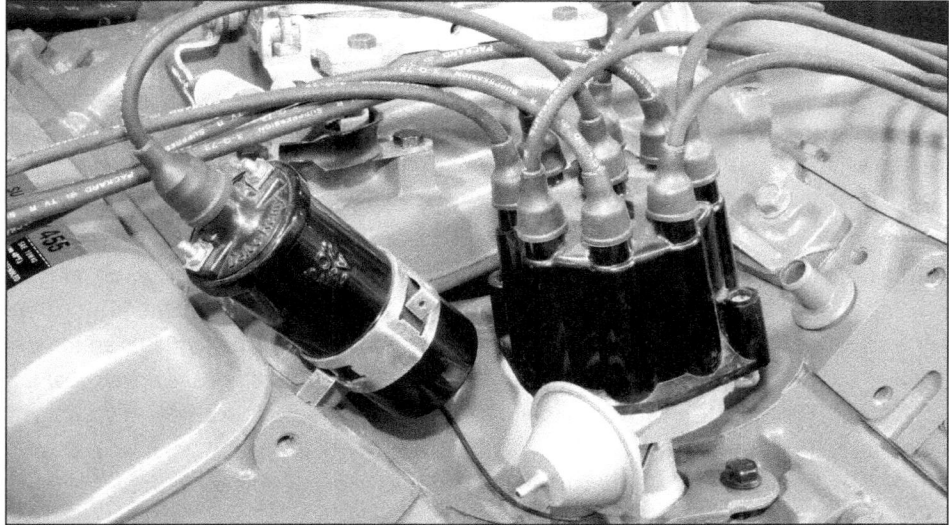

Pontiac regularly used a conventional ignition system on its V-8s produced through the 1974 model year. It consists of a separate canister-type coil and small-body distributor. Though it's connected to a 12-volt system, it sees less voltage during normal operation though a resistor wire. Though a bit of occasional maintenance is required, it's very reliable, but is considered antiquated by today's standards. Finding high-quality replacement components can be challenging.

nearly closed, and that's when it produces its greatest vacuum. The air/fuel mixture in each cylinder isn't as tightly compacted and thus isn't as combustible. Ignition must occur at an earlier crankshaft angle for combustion to consume as much fuel and air as possible, while maintaining the greatest cylinder pressure at the same ATDC crankshaft angle.

A small canister that uses engine vacuum level to sense engine load provides additional spark lead in conditions where it can benefit the engine most. The vacuum-advance canister increases spark lead by 10 to 20 degrees in light-load conditions, but it immediately relaxes as soon as engine load increases and vacuum level drops. Vacuum advance makes the engine more efficient at idle and light part throttle, and that can reduce operating temperature and improve economy. There are no negatives associated with vacuum advance and it can benefit most engines.

Points-Type System

A conventional ignition system uses a set of contact points that open and close to saturate the coil and create the arcing that crosses the plug gap to ignite the mixture. The system saturates the coil while the points are closed. The duration they remain closed is referred to as "dwell" and is expressed in degrees of crankshaft angle. Dwell has a direct effect on engine performance. A setting between 28 and 32 degrees generally provides the greatest engine performance. As the points set wears during normal use, dwell must be periodically checked and adjusted as necessary.

Beyond a bit of routine maintenance, a conventional system is very reliable, but it isn't always capable of delivering the voltage necessary for certain applications. A resistor wire reduces normal operating voltage of a conventional ignition system from 12

CHAPTER 10

Within the distributor of a conventional ignition system is a contact point set that opens and closes thousands of times each minute. The coil is saturated while the points are closed and the spark plug arcs as the points open. Because wear occurs during normal operation, the points set must be occasionally adjusted to maintain a certain amount of coil saturation, which is referred to as "dwell." I recommend a quality set like that from NAPA if replacement is required.

The HEI system is very reliable and hundreds of thousands of them were installed into Pontiac V-8s over the years. It doesn't take much hunting to find a used unit in a local salvage yard. The newest units are more than 30 years old and can be a bit cruddy from years of use. Don't let that appearance mislead you, however. This unit functioned perfectly after a thorough cleanup and proper lubrication.

If you're concerned with the reliability of a points set or are uninterested in checking and occasionally adjusting it, Pertronix offers a few different points elimination kits that use an electronic triggering system in place of the mechanical setup. The reliability of the electrical components has improved in recent years, and its Ignitor III features a built-in rev limiter that's fully adjustable.

to around 9, depending upon the application. A full 12 volts is more than the points set can handle over extended periods. Operating on less voltage reduces the amount of available voltage at the spark plug and that can make it difficult to effectively ignite the mixture in a low-compression engine, particularly with a lean mixture. This is why the system is antiquated by today's standards. Many electronic conversion kits are available.

Electronic Ignition System

Since the 1960s, Pontiac has used various ignition systems that were electronically triggered. Its Transistorized Ignition was eventually replaced by the Unitized Ignition System in the early 1970s. The mechanical contact point set in either system was replaced by an electronic pickup assembly that controlled spark firing using magnetic pulses, eliminating the need for routine maintenance

If you're planning to use a roller camshaft in your build, the distributor drive gear on the hardened-steel core isn't compatible with the iron driven gear commonly found on original and aftermarket distributors. MSD offers a high-quality bronze driven gear for a stock-diameter main shaft (.491 inch) and with a .500-inch diameter for its aftermarket distributors. The composite driven gear from BOP Engineering is available in the same sizes. Check the gear's condition seasonally, as some wear is possible.

associated with contact points. The Unitized system was self-contained and required only a single 12-volt power lead for normal operation. Long-term reliability was its weakness, but when operating as intended it was quite effective.

HOW TO BUILD MAX-PERFORMANCE PONTIAC V-8s

IGNITION

General Motors introduced its high-energy ignition (HEI) system in 1974 and Pontiac used it on a number of its V-8s beginning in the middle of the model year. The HEI system was similar to the Unitized system in that it used an electronic control module and pickup coil assembly to saturate the internal coil and trigger plug firing. Coil output was much improved and its large-diameter cap prevented cross arcing. It proved to be a very reliable ignition system and Pontiac used it on all its V-8s from 1975 forward. It can make an excellent unit, with proper preparation.

Aftermarket Ignition System

A number of companies produce new versions of the original HEI that look and function identically. The quality can vary from excellent to very poor, and the price is generally an indicator of that. If your project requires a new HEI system, I recommend those from Crane, Mallory, MSD, Performance Distributors, Pertronix, or Proform. You can expect to spend $300 or more for a new HEI system from any such company. I strongly suggest avoiding budget offerings. I have witnessed significant quality issues and would never use one on my own engine.

MSD has been producing external ignition control boxes for several years. It fires the spark plugs in each cylinder multiple times per crankshaft revolution at low to moderate engine speeds, and one very intense spark at speeds above that.

The digital processor provides constant and reliable operation at all engine speeds for extended periods. A soft-touch rev limiter is available that's designed to prevent spark from occurring at alternating cylinders, limiting the engine's top operating speed to a preset RPM. A separate box that offers many of the same options, but digitally retards spark timing is also available. A basic MSD box sells for less than $200. I have found that all available functions perform as intended.

In addition to its popular ignition-control boxes, MSD offers two popular distributors. Its Pro-Billet model features billet-aluminum construction and a hardened main shaft that rides on ball bearings. Its centrifugal advance mechanism is fully adjustable and can be locked for use in race applications with or without a crank trigger, or when a digital-retard box is used. This is an excellent distributor that sells for about $300. It does, however, require the use of an MSD box and canister coil.

MSD offers a few different ignition control boxes that use a digital processor to accurately and reliably fire the spark plugs multiple times at low to moderate engine speeds. That improves idle quality and low-speed street manners, particularly in applications where a radical camshaft dilutes the incoming mixture with residual exhaust gas. The 6AL box contains an adjustable rev limiter to prevent engine damage from operating past its redline.

Many companies produce brand-new HEI distributors for the Pontiac V-8. A unit like this from Pertronix offers top-quality construction and internal electronics. The aftermarket unit fits and functions just like an original, and it even accepts the factory wiring. You might find yourself time and money ahead purchasing such an example as opposed to hunting for a used original in excellent condition.

Pertronix offers a stand-alone small-body electronic ignition system that simply requires a 12-volt power source and canister type coil for normal operation. It's ideal for applications that are limited by intake manifold choice and/or firewall clearance. A traditional large-body HEI system isn't compatible with an original Tri-Power intake manifold. A unit like this may be the answer.

HOW TO BUILD MAX-PERFORMANCE PONTIAC V-8s

CHAPTER 10

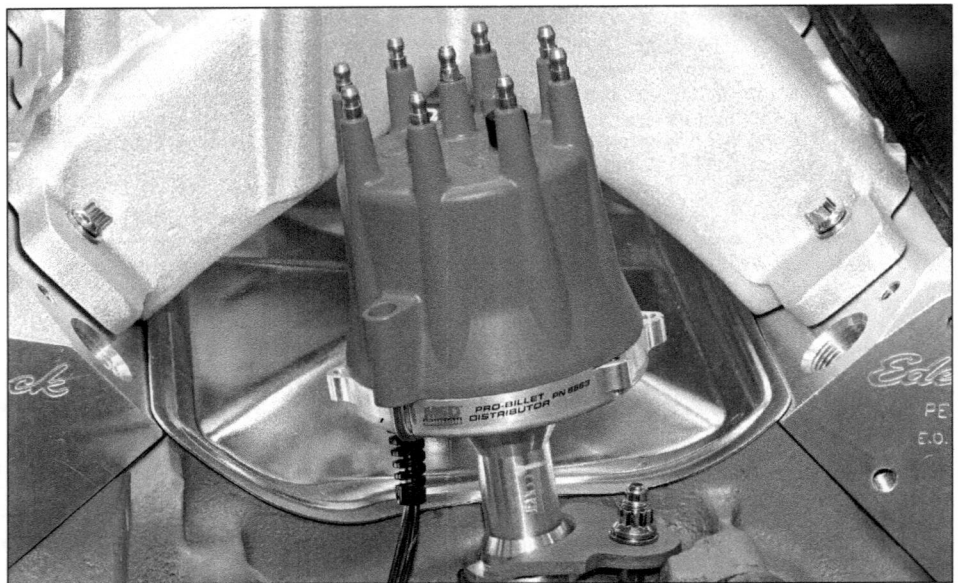

The Pro-Billet distributor from MSD is a beautifully designed unit that features a housing CNC machined from billet aluminum. The main shaft is larger than stock and floats on sealed roller bearings, and the internal electronics are of the highest quality. It requires a separate coil and ignition control box for normal operation. It's an excellent choice for any high-performance engine.

MSD offers a Pro-Billet–style Ready-To-Run distributor that contains all the same design features as the original, but has an internal module that offers stand-alone operation and includes a vacuum advance canister, which can improve its effectiveness for street-driven applications. As part of MSD's Ready-To-Run line, it's an excellent distributor and one that I use on my own Pontiac. Expect to spend about $375 for the distributor, which also requires a canister coil.

Ignition Accessories

A popular modification that's intended to improve ignition system accuracy in a competition engine is to add a crank trigger. It uses an aluminum wheel affixed to the harmonic damper to

The advance mechanism of MSD's Pro-Billet distributor uses points-type weights. It comes with various springs and limiting bushings that allow for easy advance adjustments. It wasn't uncommon to find the reluctor heavily rusted on well-used Pro-Billet distributors in the past. While it looked problematic, it didn't affect function. MSD now includes a vented cap to and a special coating on certain internals to reduce such issues.

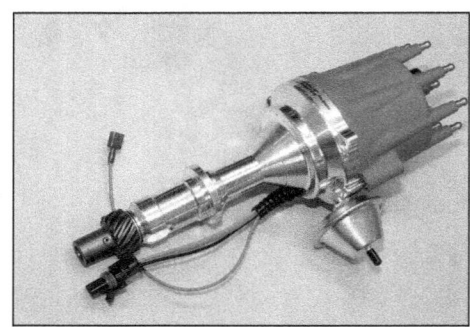

MSD offers a second Pro-Billet distributor for the Pontiac V-8 in its Ready-To-Run line. It includes an internal electronic control module and doesn't require a separate ignition control box for normal operation, but it's certainly compatible with it. A vacuum advance canister makes it an excellent choice for street applications. MSD recently added a fully adjustable internal rev limiter.

MSD's Ready-To-Run distributor contains many of the same design features as its Pro-Billet. The internal electronics are quite reliable and the advance mechanism is fully adjustable and includes the appropriate components to provide peak performance in your particular application. The Ready-To-Run unit is an excellent distributor that's a great choice when seeking to replace an ailing original.

IGNITION

sense magnetic pulses that trigger ignition firing. MSD offers an excellent crank trigger assembly, which uses high-quality electronics and a CNC-machined bracket for maximum stability and control. When using a crank trigger of any type, the distributor advance mechanism must be locked out and all timing adjustments made on the aluminum trigger wheel. The MSD assembly sells for about $300 and is compatible with its distributors, ignition control boxes and other assemblies.

Spark plug wires transfer voltage from the distributor cap to the spark plug. Any voltage leak within the wire can cause the spark to jump before it reaches the plug, causing a misfire. The stock points-type system used 7-mm wires while the HEI system used those measuring 8 mm. I regularly use top-quality 8-mm spark plugs wires on my Pontiacs and have had excellent results with spiral core wires from MSD and Taylor. Some high-voltage systems may benefit from larger-diameter wires and Taylor's Thundervolt line is an astonishing 10.4 mm thick. Taylor and MSD offer universal wire kits that must be trimmed as well as custom-fit units for Pontiac applications. Expect to spend about $100 for a complete set.

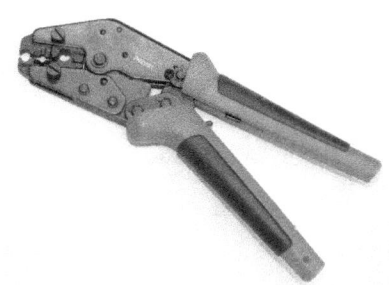

Along with a wide array of top-quality ignition components and spark plug wire sets, MSD also offers the best tool for assembling spark plugs wires that I've come across. The jaws are designed to trim the wire's outer insulation without cutting its core and the ratcheting allows for smooth and precise terminal crimping. If you prefer custom-length spark plug wires for your particular application, it's something you need in your tool box.

A crank trigger assembly improves the accuracy and stability of an ignition system, but it isn't for every application. It replaces the distributor's internal triggering components with a large wheel mounted on the harmonic balancer. The wheel contains small magnets that an electronic sensor detects, telling the ignition control box when to fire the spark plug. The crank trigger kit offered by MSD is top quality and is one I recommend.

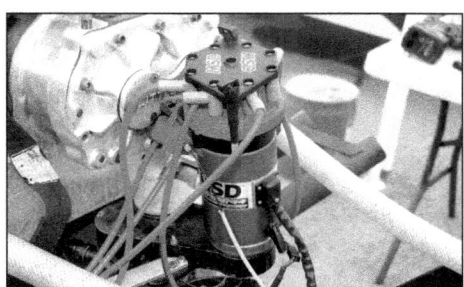

Max-performance race engines require a powerful and extremely reliable heavy-duty ignition system. To provide maximum consistency, a magneto produces its own power. That eliminates the need to rely on an external power source, such as a battery or alternator, during normal operation. MSD produces several high-quality magnetos ideal for competition engines. They are, however, a universal design and require a separate Pontiac-specific base and/or custom fitting.

Spark plugs live in a harsh environment. They must fire consistently and reliably each time, and operate at a temperature that's hot enough to prevent fouling, but not so hot that it induces detonation. Spark plugs are produced by a number of companies, but I have found those from NGK to be among the best. Available in a variety of heat ranges and models compatible with the Pontiac V-8, it's best to discuss with your spark plug requirements with your Pontiac engine builder since each application is so different.

I have been using Taylor spark plug wires in my Pontiacs for many years with excellent results. The wires are very pliable and the sockets solidly snap onto the distributor cap and spark plug terminals. Taylor's wires are available in a universal length that must be trimmed and assembled accordingly or in pre-assembled sets with appropriate lengths. They're also fully compatible with most original and aftermarket ignition systems.

CHAPTER 11

OILING SYSTEM

The oil within any engine is its lifeblood. Internal components rely on a thin film of oil to prevent metal-to-metal contact that can otherwise result in significant damage. The oil pump must be able to deliver to those components an adequate volume of oil and with sufficient pressure. The oil must be filtered, of the appropriate viscosity, and contain a sufficient level of anti-wear additives for maximum engine protection. Just how much of each is required depends upon the particular combination.

Oil

The quality of modern-spec conventional oil is about the best it's ever been. It offers excellent component protection and flows well at all temperatures. Most off-the-shelf oils are designed for modern production engines, however, and don't always contain sufficient levels of desirable anti-wear additives critical for the vintage Pontiac V-8, particularly those with flat-tappet camshafts. Because of that, I simply do not recommend modern-spec conventional oil for high-performance use.

Many large oil manufacturers produce high-quality synthetic oil that's readily available at local auto parts store, includ-

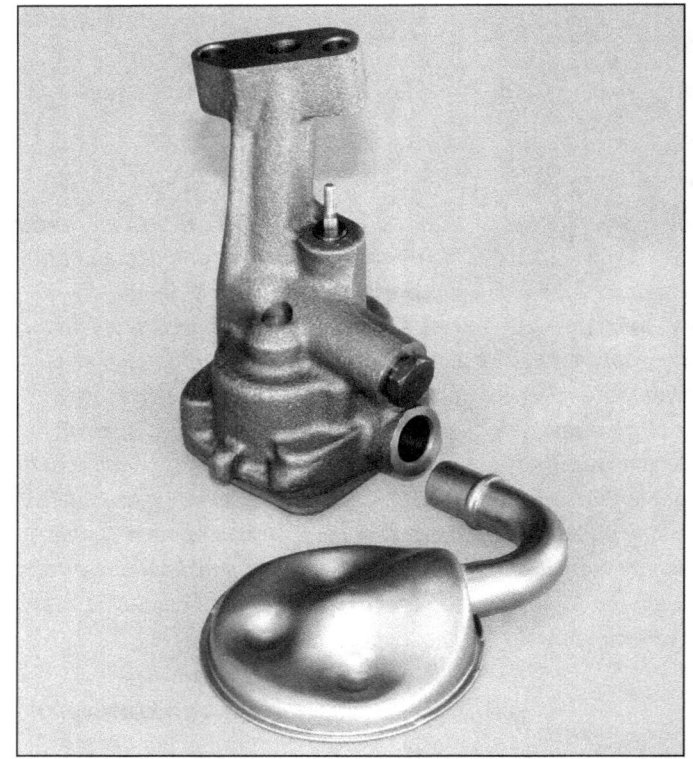

The oil pump in a Pontiac V-8 is a simple design that's connected to the distributor shaft by a long rod, and the camshaft drives them as a unit. Factory oil pumps were rarely problematic and functioned reliably in production engines for 100,000 miles or more.

ing Valvoline and Mobil 1. Synthetic oil generally flows and protects better than conventional oil, particularly at high temperatures. Modern-spec synthetic oil is suitable if you're running a roller camshaft in your Pontiac. Other oils marketed as "race" oil and/or are recommended for "off-road use only" generally have increased levels of all desirable additives and are an excellent choice regardless of camshaft type. Read the label carefully to determine its intended usage.

Heavy-duty oils designed for use in Diesel engines such as Rotella T and Delo 400 are other options. Heavy-duty oil contains high anti-wear additive levels

OILING SYSTEM

Modern oil quality is excellent but those approved for passenger car applications rarely contain sufficient levels of desirable anti-wear additives, which are required for flat-tappet use. If your Pontiac is equipped with a roller camshaft, Mobil 1 full-synthetic is an excellent choice that provides excellent lubrication and component protection.

Many professional engine builders consider Penn Grade race oil from Brad Penn among the very best available today for high-performance use. It's a partial synthetic that contains high levels of desirable additives and offers excellent protection at all performance levels. It's ideal if your Pontiac is equipped with a flat-tappet cam. Check with local machine shops or speed shops for availability. It can also be purchased through mail-order sources.

and are excellent choices that are compatible with gasoline engines. They are, however, designed for relatively low-RPM operation, and foaming at high RPM is sometimes reported. Diesel oil is a suitable oil that offers plenty of protection for a typical high-performance Pontiac V-8 operating up to 6,000 rpm, regardless of camshaft choice.

I feel the best oils available for any Pontiac V-8 are produced by Brad Penn and Joe Gibbs. These race oils are specifically designed for high-performance applications and contain high levels of all desirable additives and are compatible with flat-tappet camshafts. Available in a wide array of viscosities, both companies offer 30-weight break-in oil that's heavily concentrated and ideal for flat-tappet camshaft break in. Both brands can usually be purchased at local machine shops, speed stores, or through mail-order companies. Expect to spend $8 to $10 per quart.

Oil Viscosity

Pontiac originally specified 20- or 30-weight oil for its V-8s depending upon the season. When multi-viscosity oils became available, Pontiac revised its recommendation to include 10W-30 oil for year-round use.

While modern 10W-30 oil can provide sufficient lubrication, heavier oil is often the choice of many professional Pontiac engine builders. That's because the bearing clearances in most modern Pontiac builds are a bit looser than what Pontiac originally specified for production engines. Generally speaking, 15W-50 and 20W-50 provide sufficient oil pressure and protection. They are also a bit thicker and better resist being forced from between the contact surfaces under heavy load.

Oil Additives

Many companies produce engine oil additives designed to provide additional protection against wear. Whether specifically formulated for flat-tappet camshaft break-in or long-term use, they are simply added to the crankcase and supplement the existing oil. If you're using a

Many manufacturers offer separate additives to supplement regular engine oil. Most increase the amount of anti-wear additives in oil and/or improve lubricity. An additive isn't always required if the oil chosen is sufficiently formulated for the application, but it doesn't hurt. I regularly use GM's Engine Oil Supplement in my own engines. It's heavily concentrated and readily available from dealer parts departments.

CHAPTER 11

flat-tappet camshaft in your Pontiac, additives such as GM's Engine Oil Supplement (EOS) can prevent premature failure that's oftentimes related to additive levels in modern-spec oil. Your camshaft manufacturer or engine builders can recommend one.

Oil Pump

The Pontiac V-8 uses a positive displacement oil pump to pressurize the lubrication system and disperse oil throughout the engine. Low-performance applications received an oil pump regulated to 40 psi, while higher-performance engines generally received a 60-psi pump. A limited number of applications received an 80-psi unit. The basic design works sufficiently for the intended applications, but its main weakness is its ball-type bypass valve. Should debris get caught between the check ball and seat, excessively low oil at idle pressure can result.

Melling produces several different oil pumps for the Pontiac V-8. In addition to 60- and 80-psi pumps in its basic line, Melling offers those same pressure ratings in its Melling Select line; a higher end line that's assembled to tighter tolerances, and can require less preparation for high-performance applications.

Whenever using an out-of-box oil pump in any rebuild, I highly recommend completely disassembling the unit and closely inspecting for any metallic particles or debris that can prevent the check ball from fully seating during normal operation. Small burrs can be

Most aftermarket oil pump pickup assemblies are retained by a bolt, which prevents it from working loose during normal operation. If you're using a stock oil pan and pickup assembly in your build, I highly recommend applying a spot weld that permanently joins the two. Only a small tack is needed and as long as it's applied quickly, it shouldn't weaken or damage the cast housing.

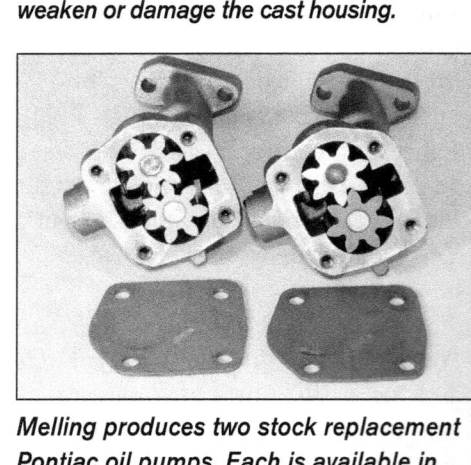

Melling produces two stock replacement Pontiac oil pumps. Each is available in 50- or 70-psi variants, which generally produces 60 and 80 psi, respectively, with normal production bearing tolerances. The standard pump (left) is built to exact OE specifications. The Select Performance pump (right) features tightened tolerances and billet gears to improve performance and reliability, particularly at high engine speed.

The Pontiac oil pump is relatively simple. Its cast body houses two spur gears that draw in, pressurize, and disperse oil throughout the engine. A ball-type pressure-relief valve regulates oil pressure to 40-, 60-, or 80-psi depending on the unit and engine build conditions. The stock oil pump is suitable for engines producing 600 hp or 6,000 rpm, or possibly more.

OILING SYSTEM

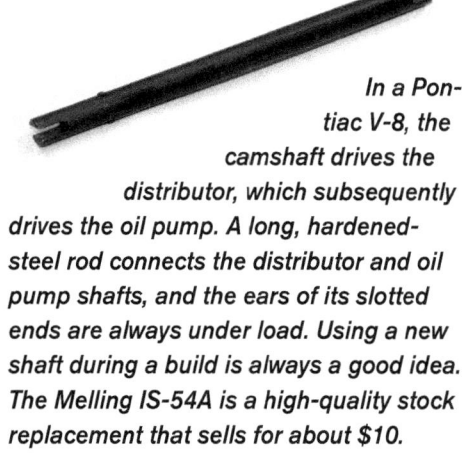

In a Pontiac V-8, the camshaft drives the distributor, which subsequently drives the oil pump. A long, hardened-steel rod connects the distributor and oil pump shafts, and the ears of its slotted ends are always under load. Using a new shaft during a build is always a good idea. The Melling IS-54A is a high-quality stock replacement that sells for about $10.

removed with a file or stone. Major issue should be pointed out to your machinist for a professional opinion. Once the oil pump is perfectly clean, it can be carefully reassembled, the pickup installed, and pressed into service. There's no reason it shouldn't perform reliably in engines operating up to 6,000 rpm.

In my opinion, the best oil pump available today is from Butler Performance (BP). Its Pro-Series pump starts life as a typical 80-psi Melling unit, but it's completely disassembled, re-machined, blueprinted, and reassembled.

During the re-machining process, lubrication grooves are added to certain areas and the idler gear bore is honed and a relief hole is added to the idler gear to prevent oil cavitation; a condition where trapped air pockets collapse, causing irregular shock loads that can prematurely fatigue components.

During reassembly BP adds a very thick bottom plate to prevent pressure loss associated with distortion. The pump is then pressure and volume tested to ensure it provides 70 to 80 psi and can flow as much as 8 to 10 gpm more than an unmodified 60- or 80-psi unit. Delivered in ready-to-run condition, no additional preparation steps are required and you can be assured that it will provide immediate oil pressure upon startup, saving time and effort versus preparing a typical unit.

The Pro-Series oil pump from Butler Performance may be the best OE-type Pontiac pump available today. It begins life as a standard Melling unit that Butler blueprints and adds a thick bottom plate to maximize performance and reliability in applications producing 1,000 hp or more. It's wet flowed and pressure tested before shipping, which reduces the chance of oil-pressure-related issues upon immediate startup.

For a cost of less than $200, the Pro-Series pump provides the reliability and consistency required for max-performance applications. It's ideal for most Pontiac builds producing 550 hp or more, and is very capable of providing sufficient lubrication for engines producing 1,000 to 1,200 hp, or possibly even more. Specialty gerotor or billet dry sump oil pumps may be required for applications beyond 1,200 hp. They are, however, expensive and are generally a custom-order setup. I recommended contacting Butler Performance for more oil pump information if you feel your particular engine requires lubrication beyond the capability of its Pro-Series pump.

The internal gear surfaces within the Butler Performance Pro-Series oil pump are diamond-lapped for improved consistency. The drive gear shaft is micro-polished and positive lubrication is added to reduce friction and wear. The check ball in the bypass valve and its sealing surface are also blueprinted for maximum pressure consistency.

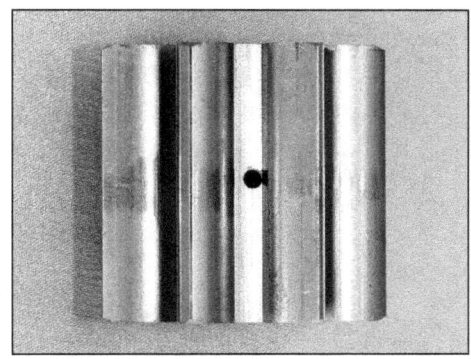

A hole is added to the driven gear to bleed off trapped air, which leads to cavitation. It's a condition where trapped air pockets collapse during operation, creating irregular shock loads that causes component fatigue and failure. A groove machined into the bottom plate works in conjunction with the modified gear to further reduce the chance of cavitation.

Oil Pressure Requirements

Oil pressure provides a constant flow of oil to the bearings, and generally speaking, 10 psi of oil pressure for every 1,000 rpm is the minimum requirement for reliable operation. Ideal hot-idle oil pressure should be between 30 and 40 psi when using an oil viscosity of 30W or greater. Pressure should climb quickly as

CHAPTER 11

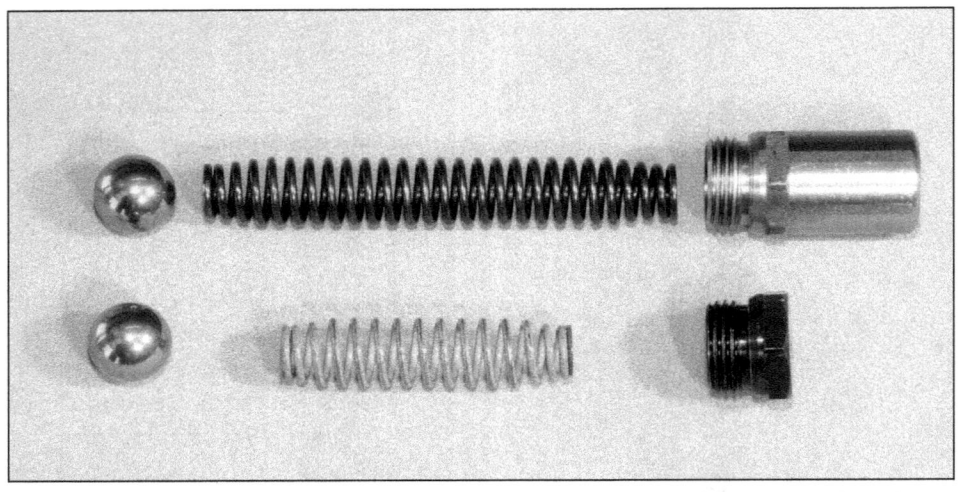

The Pontiac oil pumps available today use the same cast body while the length and tension of the bypass valvespring regulates pressure. The shorter spring provides about 60 psi and the longer spring provides about 80 psi. Installing a .060-inch shim between the spring and cap is a common modification in applications where a slight pressure increase is needed.

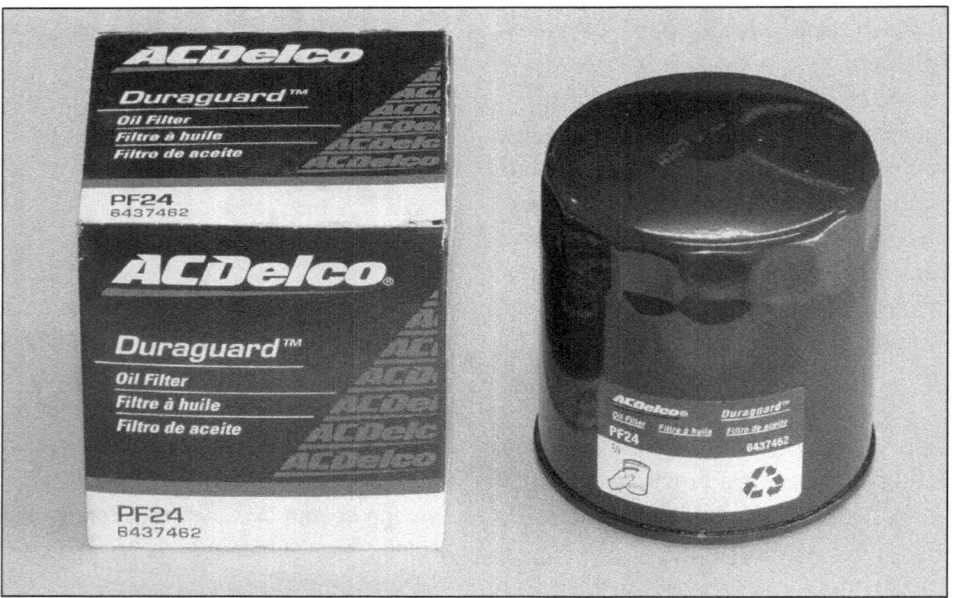

The PF24 oil filter from AC Delco is a stock replacement specifically designed for the Pontiac V-8. It's a high-quality unit that offers excellent filtration and can generally withstand high oil pressure levels. It can be ordered through most local auto parts stores if it's not already on the shelf.

engine speed increases, and it should top out between 60 and 80 psi in any high-performance application.

An oil pump with a pressure regulator setting of 60 psi is generally adequate for most applications. Some argue that pressure greater than that does little more than create parasitic drag that reduces horsepower. With slightly greater bearing clearances typically associated with modern high-performance builds, an oil pump with an 80-psi regulator setting might be a better choice as it generates only 70 to 75 psi within the engine.

Oil pump pressure may be a consideration when using a very aggressive hydraulic-roller camshaft in a high-performance build. The high-pressure valvesprings required to effectively maintain valvetrain stability can cause the roller lifter's hydraulic internals to bleed down too quickly, and that can translate into a significant performance loss. In those instances, the 60 psi associated with a typical oil pump may not be enough. Additional oil pressure may be required to keep the hydraulic lifters functioning properly.

Oil Filter

An oil filter is designed to remove and collect small particles and debris that's suspended in oil as it circulates throughout the engine. AC provided Pontiac with its oil filters for its V-8s. It's a high-quality design that's rather large and offered excellent filtration and flow. AC Delco has since assumed production and its number-PF24 filter is a modern replacement for the Pontiac V-8. It's not commonly stocked on local auto parts store shelves, but it's something that can be ordered in and arrive within one day.

I have found that the Pontiac-spec filter sold by NAPA in its Gold line is produced by Wix. Number-1258 is a high-quality filter that I've often substituted for the AC Delco PF-24 without issue. Mobil 1 also produces high-quality oil filters, and its M1-203 is considered a stock replacement for the Pontiac V-8. Any of these filters would make an excellent choice for your Pontiac V-8.

A remote filter assembly can be used if larger-than-stock filter capacity is needed or if exhaust header routing prevents the use of an appropriately sized filter. I recommend discussing remote filter options with your builder.

OILING SYSTEM

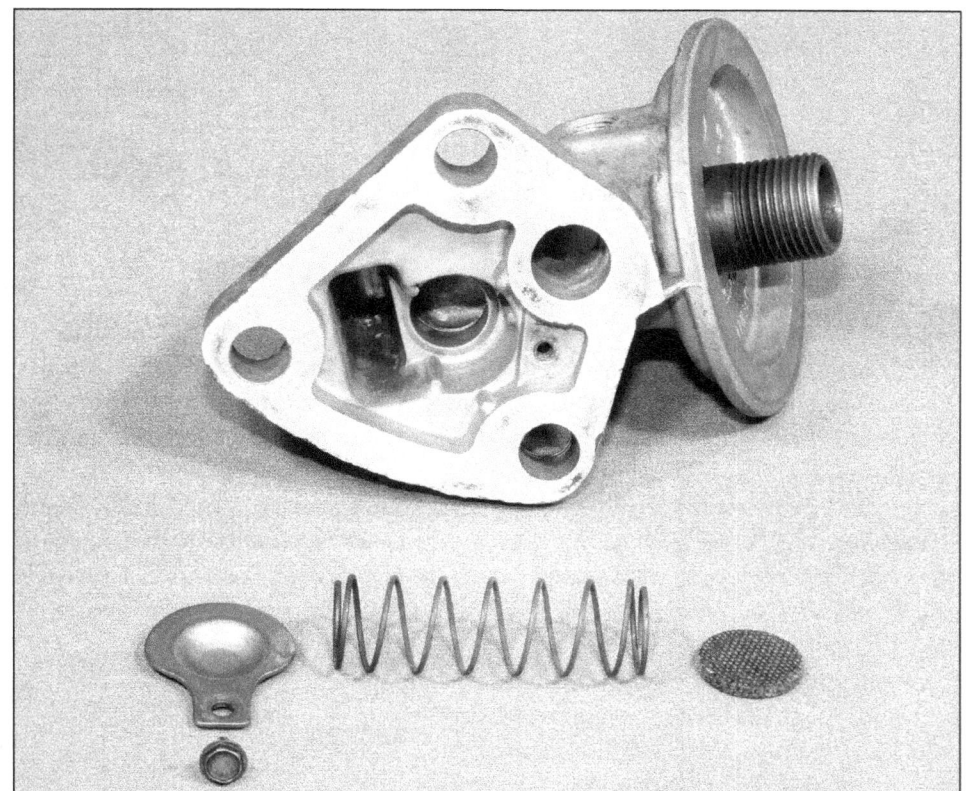

The bypass in the stock oil filter housing allows for continuous oil circulation should filter flow drastically decrease or stop. Performance enthusiasts commonly block the bypass, forcing all the oil through the filter, preventing the chance of circulating unfiltered oil. Strong arguments can be presented for either case, but I prefer to leave it functional since I believe no oil flow is a greater risk to the engine than unfiltered oil.

APE and KRE reproduce Pontiac oil filter housing. The unit from APE includes a stock-type bypass while the unit from KRE is left as cast, and not functional. Both are distributed by many companies for less than $75.

It usually consists of a remote mount, flange adapter, and the required high-pressure hoses.

Oil Filter Housing

Pontiac included an internal bypass in its oil filter housing that's designed to let pressurized oil bypass the filter should there ever be an impedance to flow. It prevents the bearing and journal damage that can occur from lack of oil flow if the filter ever fails. In fear of unfiltered oil circulating throughout their engine, many hobbyists block the bypass by inserting a pipe plug into the passage to force 100 percent of the oil through the filter. I don't see unfiltered oil as a critical issue, however, especially if the bypass is operating as intended.

The bypass operates on a pressure differential and if there's a momentary pressure loss of the oil exiting the filter, the bypass allows a slight bit to slip past to sufficiently lubricate the bearings. With regular oil changes and quality filtration it's unlikely that the oil is dirty enough to cause significant damage. I'd rather have unfiltered oil flowing throughout my engine that no oil at all.

If a filter issue ever arises, a teardown is likely in the immediate future anyway, and if the bearings are performing their intended task, the foreign particles suspended in the unfiltered oil should imbed into the bearing as opposed to scoring the journal surfaces. That can leave the components in good, reusable condition as opposed to being severely damaged if no oil flow occurs.

I recommend simply disassembling the oil filter housing bypass, thoroughly cleaning it, and then reassembling it. You certainly can, however, tap the housing with an appropriately sized pipe plug if you're concerned with the chance of unfiltered oil circulating in your engine. There are several companies reproducing factory oil filter housings. Some contain a factory-type bypass and others are cast shut.

Oil Pan

A typical Pontiac oil pan has a large sump at the rear that the oil flows into and the pump draws from. In race-type applications, the very best pan keeps the oil from climbing up the back wall, and uncovering the pump during very hard acceleration. Curing this can require a wider or deeper sump to increase the capacity. Professional Pontiac engine builders have their favorites, and they are usually custom units. You can discuss this with your Pontiac engine builder if you feel a drag-race oil pan is required.

There are also a few readily available aftermarket oil pans ideal for

CHAPTER 11

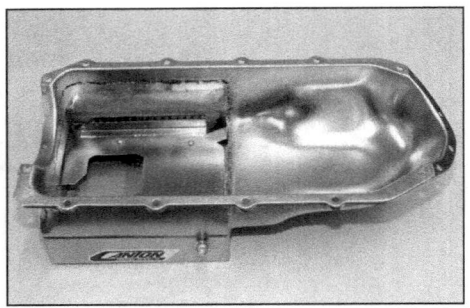

Canton produces several oil pans for the Pontiac V-8 that ranges from stock replacement to high-performance street/strip units with additional capacity. Expect high-quality construction and excellent fit and finish. It even offers variants compatible with most popular Pontiac models. Its Road Race unit (shown) includes trap doors and an integral baffle that doubles as a windage tray.

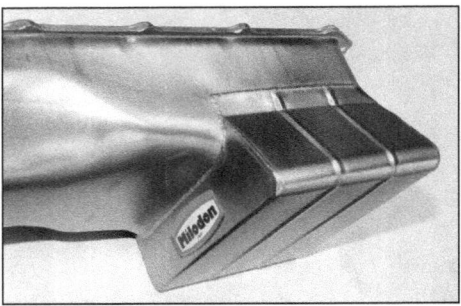

Milodon produces Pontiac oil pans ranging from a stock replacement-type to a road race unit complete with trap doors. Its low-profile oil pan fits and functions like an original, but offers additional capacity with a unique kick-out feature. It's a very popular unit that fits most Pontiac models. Because it's about 1 inch deeper than a stock unit, a specific pickup is necessary.

If you plan on using the factory oil dipstick in your Pontiac, be sure it and its tubing are compatible with the windage tray and oil pan if you're purchasing aftermarket pieces. The factory tray also retained the lower dipstick tube. If you're using it or a Tomahawk replacement and your original tube is missing, a high-quality replacement is available from Pontiac vendors.

Unbolting the removable baffle reveals the trap door system that makes the Canton Road Race oil pan so effective. The doors are designed and positioned to allow easy oil flow toward the oil pump pickup during acceleration, and prevents it from flowing away during hard cornering or braking. It works in conjunction with a specific pickup assembly that Canton sells separately.

Milodon produces a few different oil pans for popular Pontiac applications, and they seem popular with hobbyists. Its low-profile pan includes a kick-out feature that increases oil capacity and has internal baffling helps keep the pickup submerged in oil during harsh maneuvers. Those I've spoken with report excellent fit and function. It requires the use of a specific oil pump pickup, which Milodon offers. Expect to spend about $300 for any of the offerings.

A dry sump oil system removes oil that collects in the oil pan sump and relocates it to an external canister where it's dispersed throughout the engine by an external oil pump. While not practical for every application, a dry sump system offers a few distinct advantages in max-performance engines, inlcuding increased lubrication system capacity and potentially greater engine output as it essentially eliminates the windage present in a typical wet sump engine.

The main disadvantage to a dry sump setup is cost. I know of no company presently mass producing complete dry sump kits for the Pontiac V-8. That means modifying certain components

Pontiac eliminated the windage tray on its V-8s during the mid 1970s. The lower dipstick tube was retained by a bracket that simply bolted to the number-3 main bearing cap in those applications. If you're using an aftermarket windage tray in your particular build, a tube like this may be an option that allows the use of original oil dipstick. You can source a replacement from many Pontiac vendors.

high-performance use. Canton's Road Race oil pan for GTO and Firebird models uses an internal baffle and trap doors to keep the sump full during hard braking, accelerating, and hard cornering. The wider-than-stock design increases the pan's capacity to 7 quarts or more. It fits very well and functions as intended. The internal baffle must be drilled if you plan to use the factory oil dipstick, so some test fitting may be necessary. It sells for about $300.

for other makes to fit or newly fabricating others, and that can drive up the cost on an already expensive system. If you feel a dry sump oiling system may be something your Pontiac requires, your engine builder can recommend the best components for you. Expect to spend several thousand dollars to convert your Pontiac V-8 to dry sump, however.

CHAPTER 12

TUNING

Any engine must be tuned optimally for it to produce as much horsepower and torque as it possibly can. That includes adjusting its carburetor and distributor until the fuel and spark curves are perfect at every possible speed. Achieving that can be an arduous and time-consuming process with much trial and error, but the efforts can significantly improve a vehicle's performance.

Carburetor Tuning

A carburetor is responsible for providing the engine with the optimal amount of fuel to provide peak performance in all operating conditions. Once you've determined how much airflow capacity your engine requires and purchased the appropriate carburetor, you must begin to adjust the internal fuel metering to meet the demands of your particular engine. That can be a time-consuming process depending upon the carburetor.

The Rochester Quadrajet remains popular with Pontiac hobbyists because it was the factory-installed 4-barrel in most popular applications, is compatible with original equipment, and is relatively inexpensive. Its small primary barrels offer excellent throttle response and fuel economy and its larger secondary barrels are vacuum actuated and supply fuel as demand requires. It's an ideal carburetor for street drivers and can make an excellent street/strip carb so long as the vehicle's fuel system is capable of keeping the centrally located float bowl full of fuel.

Pontiac used the Quadrajet on engines ranging from 301 to 455 ci.

Engine tuning is a time-consuming process that's required to achieve peak performance. It generally includes swapping carburetor jets and adjusting the distributor to advance or retard spark timing. The only way to determine optimal tune is to measure engine output on the race track or with a dynamometer. Anything else is just a guess.

When modified properly, even a casting originally installed onto a lowly 301 can make an excellent performance unit for Pontiacs displacing 470 ci or more. The required modifications include drilling out various passages to the appropriate sizes, and installing new metering jets and rods. CarTech's book *How to Rebuild and Modify Rochester Quadrajet Carburetors*

CHAPTER 12

The Rochester Quadrajet is an excellent carburetor. Those from the mid-to-late 1970s offer a maximum airflow capacity of 800 cfm or slightly more and are the best performance candidates for the cost. Originally installed on Pontiac V-8s displacing 301 to 455 ci, they all respond similarly and favorably to performance modifications regardless of original application.

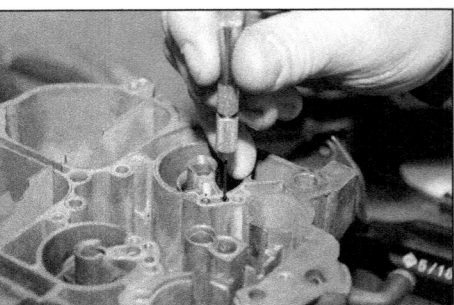

To achieve maximum possible performance from any Quadrajet casting, it must be modified accordingly, and that includes drilling out internal passages to optimize emulsification in certain conditions. A small machinist's drill-bit set and a pin vise is required for absolute precision. You'll be amazed at how well a Quadrajet performs with a few simple adjustments!

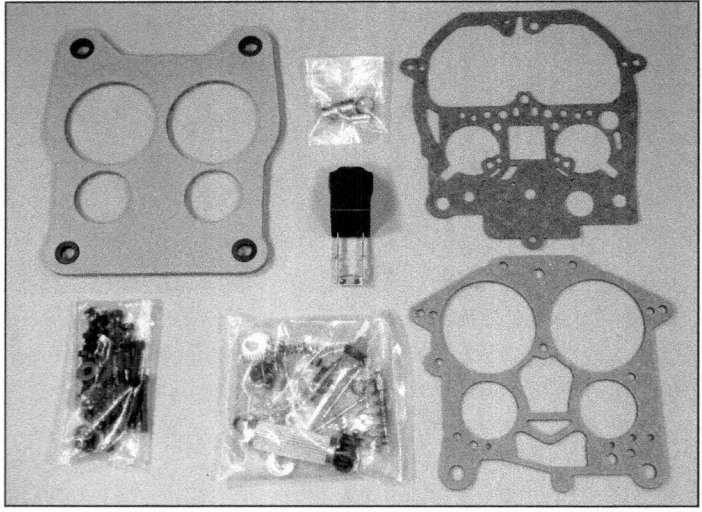

Sourcing high-quality Quadrajet rebuild components from local parts stores can be difficult. I recommend purchasing from companies that specialize in Quadrajet repair and recalibration. Cliff's High Performance offers complete rebuild kits containing the highest quality components available, some even made to Cliff's specifications. Its kits are completely compatible with Ethanol-blend fuel as well.

by Cliff Ruggles is an excellent resource that clearly explains how best to modify your Quadrajet for a particular application as well as ideal metering jet and rod staring points.

It can be difficult to find high-quality replacement components for your Quadrajet from local auto parts stores. Rebuild kits are often rather generic and the rubber components are not compatible with the ethanol-blended fuels so common today. I recommend sourcing your parts from Cliff's High Performance. Its rebuild kits include top-quality components that are impervious to ethanol-blend fuel. Cliff's also offers new jets and rods that allow you to finely tune your Quadrajet's fuel curve for peak performance in all conditions.

The Holley 4-barrel is the most popular aftermarket carburetor for high-performance use. A great number have been produced over the years, and finding a new unit with the appropriate airflow capacity for your particular engine is as easy as opening the catalog from a mail-order supplier. The Holley 4-barrel remains very popular because it's easily adjustable, offers two large float bowls, and can accurately and consistently provide the engine with a constant supply of fuel to achieve peak performance.

The Holley HP series carburetor includes removable air jets to fine tune idle and high-speed mixing as well as easily accessible fuel jets in the metering blocks. The Ultra HP series offers even greater adjustability. I have found it best to start with Holley's out-of-box fuel

TUNING

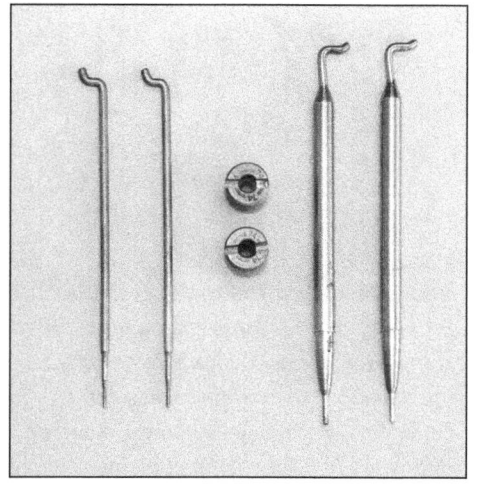

Holley 4-barrels are popular for high-performance use. Many of the modifications once routinely performed to improve the performance of older Holley units are now standard features on the HP and Ultra HP models. The adjustability makes them very tunable. Capable of delivering stellar performance and excellent street manners, they're very popular with hobbyists.

A Quadrajet uses removable jets and metering rods for primary fueling and removable secondary rods with fixed jets for secondary fueling. A wide array of sizes was available through Delco for years, but most are no longer available. Junkyard scouring can provide a small assortment, but it may not be enough to satisfy your engine's fuel needs. Cliff's High Performance provides most jet or rod sizes you might need.

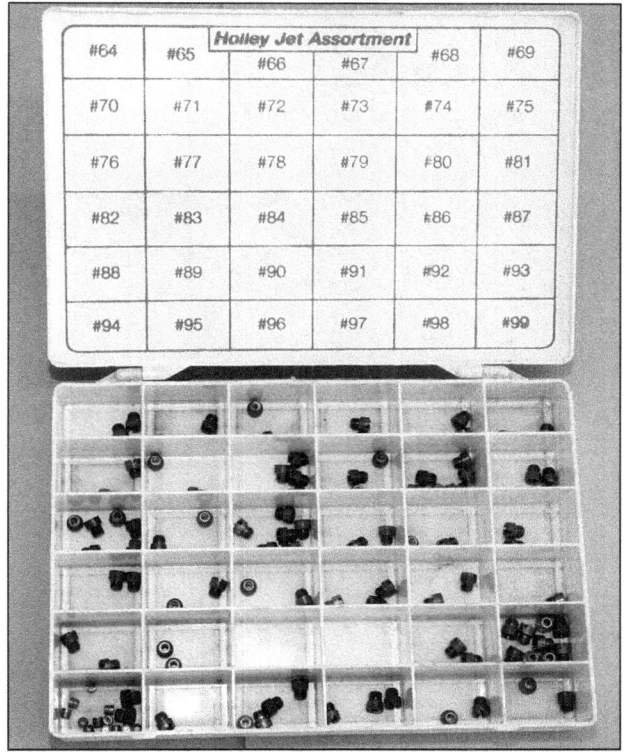

Holley offers complete jet kits to finely tune its carburetors for the exact application. New sets are available from mail-order suppliers and cost $100 or more. I found this kit at a swap meet for $20. It was missing only a few jets sizes, which I found at a local speed store, completing my assortment.

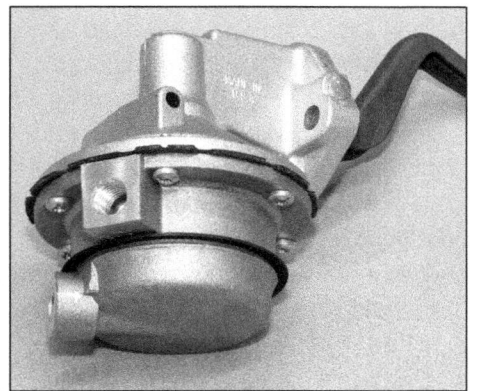

If you're using a mechanical fuel pump in your Pontiac V-8, a high-performance unit from Carter or RobbMc Performance may be required to deliver sufficient volume for engines producing 500 hp or slightly more. I suggest an electric fuel pump once horsepower exceeds that point, particularly if you're using a Quadrajet with its very small float bowl. A block-off plate can be installed in place of the mechanical pump when using an electric unit.

The HP and Ultra HP carburetors from Holley feature screw-in idle and high-speed air jets, which allow for emulsification tuning in various conditions. Holley sells these small jets in individual pairs or complete tuning kits. It's an excellent way to maximize the performance of your engine when using such a carburetor. You may need to consult with a professional tuner before deviating from Holley's base setting.

CHAPTER 12

The metering block of the new Ultra HP Holley carburetor is CNC machined from billet aluminum and its features removable emulsification jets. Adjusting jet size has a dramatic effect on performance and simply swapping them around creates other operating issues. I suggest consulting a Holley tuning specialist to determine the direction that best suits your Pontiac.

Professional distributor tuning equipment is the easiest way to measure and adjust the advance curve within a distributor. Quite common many years ago, it can be difficult to find functional equipment today. Some machine shops still offer this service, but serious racers generally know a private individual who owns one, so ask around.

settings and adjust accordingly. Chances are it will be a bit rich in all conditions, but it's best to establish a baseline to determine the direction you need to go. Holley offers complete tuning kits that include several air and fuel jets, and various power valves are also available.

Distributor Tuning

Maximum possible torque is achieved when peak combustion pressure occurs at a crankshaft angle 10 to 12 degrees ATDC in all conditions. The distributor must be timed appropriately to provide peak performance and that includes initiating combustion some 30 to 40 degrees before the piston reaches TDC, depending upon the application. It's not always practical to run that much spark advance in a street engine at all times, so the distributor uses a mechanical advance mechanism that allows a lower base timing for easy starting, and increases advance as engine speed climbs.

While electronically controlled units (such as those available with some fuel injection kits) respond to electronic input, a typical distributor must be manu-

Jet changes on a Holley carburetor are rather simple. It's as easy as removing the four float-bowl retaining screws and setting the bowl aside, and then unscrewing the jets from the metering block. I recommend starting with Holley's box-stock setting and adjusting accordingly once you've determined if your Pontiac's fuel curve needs to be richened or leaned.

ally adjusted, and proper adjustment generally requires swapping parts and measuring the effects using a dial-back timing light or a professional distributor testing equipment. Bob Davis Distributors offers "recurve" kits that allow for easy and accurate adjustments of factory equipment. Some aftermarket distributors accept them too. High-quality distributors, such as those from MSD include a kit to accurately adjust the advance curve.

MSD distributors include an adjustable advance mechanism that uses high-quality springs and limiting bushings to control how much advance is available and the rate at which it comes in. It's an excellent design that allows for easy and accurate adjustments. MSD also offers its tuning kit separately should you need replacements.

For a naturally aspirated Pontiac V-8, I recommend a base (or initial) timing anywhere between 10 and 15 degrees. A mechanical advance curve that provides 20 to 24 degrees of additional advance can provide the optimal amount of total spark lead required for peak performance. Some trial and error will be required to find the exact amount your Pontiac runs best with. If at any time during your carburetor and distributor tuning process you hear detonation at full throttle, let off

TUNING

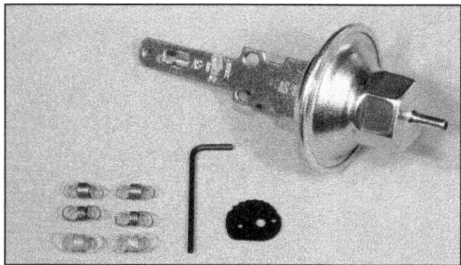

The Adjustable Vacuum Advance kit from Crane includes a variety of advance-curve springs as well as a serrated limited plate and vacuum adjustment tool. It's an excellent piece that's ideal for tailoring vacuum advance to provide peak engine efficiency at idle and part throttle. I recommend it above all others.

Dyno tuning is an easy and accurate way to ensure that your Pontiac is running its best. An engine dyno uses a water brake to measure engine output. Timing and carburetor adjustments are easily performed and the performance effects can quickly be measured. A slight amount of tune adjustment may be required to compensate for vehicle weight once the engine is installed.

immediately or significant engine damage can result. You may need to retard spark timing or richen the fuel mixture, or possibly a combination of the two.

All distributor tuning should be performed without the vacuum advance canister connected, if so equipped, since it has no bearing on full-throttle performance. Once the appropriate amount of base and total spark lead for peak output is determined, connecting the advance canister to a manifold vacuum source provides additional spark advance at idle and low-speed conditions. Crane offers an excellent vacuum advance kit that includes a fully adjustable canister. I recommend it if you plan to use vacuum advance with your combination.

I suggest starting with a setting that provides about 12 degrees of vacuum advance at vacuum setting that's higher than what your engine produces at idle speed. Connecting the canister to a manifold vacuum source provides vacuum advance in all conditions where sufficient vacuum is present. That includes idle, light acceleration, and deceleration. If you detect surging under part-throttle load, reduce the amount of vacuum advance in 2-degree increments until it subsides. If you detect surging while decelerating, simply connect the vacuum can to a ported vacuum source.

Dyno Tuning

A dynamometer measures engine output under full-throttle conditions. There are two dyno types: engine and chassis. Each can provide you with valuable information about your engine's state of tune, and the effects that fuel and spark adjustments have on it. I strongly recommend having your engine dyno tuned, as it provides the greatest possible performance from your particular combination.

Nothing replaces actually dyno testing your engine, but quality computer simulators like that from Performance Trends predict engine output with reasonable accuracy. I found it to be an excellent tool when attempting to understand the effects certain components have on performance. I regularly use the Engine Analyzer v3.4 program when planning complete builds or modifications to existing ones.

HOW TO BUILD MAX-PERFORMANCE PONTIAC V-8s

CHAPTER 12

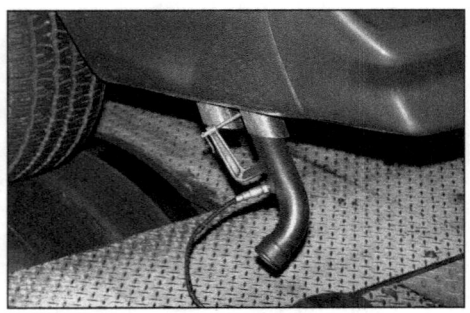

A wideband air/fuel (A/F) meter is an excellent way to accurate determine the air/fuel ratio of your particular engine and measure and record the effects of your tuning efforts. An A/F ratio near 12.7:1 generally produces peak full-throttle performance while a ratio of 14.7:1 or greater is ideal for idle and part throttle. Innovate Motorsports makes an excellent A/F meter that's ideal for home hobbyist use.

A chassis dyno measures engine output through the driveline. It requires that the engine be installed and in good running condition. Because of driveline loss, the results generated using a chassis dyno are generally lower than those recorded on an engine dyno. A mathematical equation provides approximate crankshaft horsepower and torque figures from those measured on a chassis dyno.

The roller drum of some chassis dyno models contains a fixed weight while others use an eddy-brake that adds rolling resistance to replicate the load the engine sees accelerating the vehicle down the road or race track. That is very beneficial in achieving maximum performance from your engine.

An engine dyno measures power output through a water brake connected to the crankshaft. Numbers recorded using an engine dyno are referred to as "at the crank." A chassis dyno measures engine output of a running vehicle by positioning the drive wheels on a large roller drum and measuring how quickly the engine accelerates against the resistance of the roller drum. Chassis dyno results are commonly referred to as "at the tire," and the horsepower and torque values are often much lower because of driveline loss. A mathematical conversion can be used to approximate numbers at the crankshaft.

Various sensors are typically connected to the engine during a dyno tuning session. The most important of which is a wideband air/fuel sensor that records the air/fuel (A/F) ratio the engine is emitting from its exhaust. A ratio of 14.7:1 is stoichiometric (or perfect) but it rarely yields greatest full-throttle performance in a naturally aspirated Pontiac. An average A/F ratio that's slightly richer and usually closer to 12.7:1 tends to produce greatest power output. There are a great number of variables that can affect it and the dyno results reveal the best exact A/F ratio for your Pontiac.

Volumetric Efficiency

Volumetric efficiency (VE) is a measure of how effectively an engine can ingest and expel air at full throttle during one complete cycle. It is generally stated as a ratio or percentage that's calculated by dividing actual airflow rate during normal operation by the engine's maximum amount of static volume.

VE is used to determine how efficiently an engine operates at a given RPM. While it can be mathematically calculated using a combination of known values and estimations and a rather complicated formula, the easiest way to accurately determine the VE of a particular engine is likely with an engine dyno. High-end dynos use a mass airflow meter to accurately calculate VE at every point throughout the entire pull by comparing the measured voluminous amount of ingested airflow to actual displacement.

Peaking at about the same point as torque output, a relatively low VE can indicate a significant rotating friction or a restriction within the intake or exhaust tract, preventing an engine from reaching its full pumping potential. Components aimed at reducing frictional loss include short-skirt pistons with low-tension piston rings, a low-mass roller valvetrain, and a lightweight rotating assembly or improving intake or exhaust airflow of a particular engine. These usually improve VE, and that can result in greater total power output, but particularly at high engine speed.

A larger engine has a greater swept volume, and it can displace a greater amount of air with each stroke, so at low speed, its VE is generally greater than that of a smaller engine at a particular RPM. A smaller engine can achieve as much or more VE, but since it doesn't displace as much volume with each stroke, it must operate at higher engine RPM to achieve the same amount of voluminous flow. That speed is sometimes impractical for the heavier reciprocating mass associated with large-cube engines.

A VE greater than 90 percent from a naturally aspirated Pontiac V-8 is considered

TUNING

A vacuum pump is used to create negative pressure within the crankcase, reducing pumping loss and oil leaks while improving cylinder seal, all of which improve performance. Moroso offers an excellent vacuum pump kit that's the choice of many professional Pontiac engine builders. While Moroso doesn't offer the appropriate mounting brackets and drive equipment to allow easy installation onto your Pontiac V-8, your favorite Pontiac builder likely does.

A wideband air/fuel (A/F) meter is an excellent way to accurate determine the air/fuel ratio of your particular engine and measure and record the effects of your tuning efforts. An A/F ratio near 12.7:1 generally produces peak full-throttle performance while a ratio of 14.7:1 or greater is ideal for idle and part throttle. Innovate Motorsports makes an excellent A/F meter that's ideal for home hobbyist use. (Photo Courtesy Nitrous Oxide Systems)

excellent and tuning for a VE as close to 100 percent as possible usually yields the greatest performance. A VE of more than 100 percent can occur in certain ranges with a properly designed camshaft and/or exhaust system that scavenges the cylinders, drawing in the intake charge during the valve overlap period. Since induction is forced into the cylinders of a turbocharged or supercharged engine, it's not uncommon to find VE greater than 100 percent at times. Your dyno operator can likely provide your engine's VE, and you can discuss it with your engine builder to determine if it's a suitable amount for your combination.

Vacuum Pump

An engine can generate a significant amount of internal windage as the crankshaft rotates and the connecting rods churn about. A positive crankcase ventilation (PCV) system is a factory-designed concept that keeps oil mist and gas fumes from exiting the crankcase, but its effectiveness in reducing internal pressure is limited. A vacuum pump can relieve an engine of any internal pressure, allowing for quicker engine acceleration, the use of low tension oil rings to reduce friction, and the reduction of oil leaks.

The vacuum pump is mounted onto the engine using an adjustable bracket; it's driven by a belt connected to the crankshaft. The amount of vacuum the pump generates increases along with engine speed, and it draws though oil-resistant hoses affixed to specially modified valve covers; the route is to a large filter where the oil and air are separated. By creating a slight depression within the crankcase, the pistons are no longer working against internal pressure when traveling downward in the bore. It also allows the piston rings to better seal against the walls, and that can reduce the amount of frictional loss associated with high tension and/or thicker ring packs.

If there's available space in your Pontiac's engine compartment, there are no negatives associated with the addition of a vacuum pump. It's not uncommon to find horsepower increases of 10 to 30 depending upon the application.

Moroso makes an excellent assembly and many professional Pontiac engine builders offer specific brackets, crankshaft mandrels, and pulleys for easy installation onto a Pontiac V-8. Expect to spend $800 or more for a complete kit, and if you plan to use a lip-type rear main seal, you might consider installing it backward, which improves its sealing ability when under vacuum.

Nitrous Oxide

Using nitrous oxide to improve engine performance has been a popular modification for many decades. It involves injecting high-pressure oxygen molecules into the engine, where it can enter the cylinders providing a denser charge when the nitrogen and oxygen molecules separate when exposed to extreme heat. By providing an additional amount of fuel that's proportionate with

CHAPTER 12

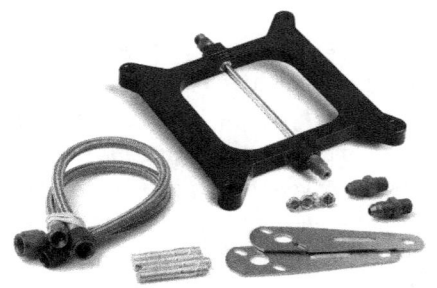

Many hobbyists modify their intake manifold by drilling ports into each runner to install small nitrous injectors. That provides an equal amount of nitrous oxide to each cylinder. NOS also produces a variety of spacer plates that fit between the carburetor and intake manifold, delivering nitrous oxide directly to the plenum. It's an excellent method that delivers nearly as much performance as the injectors, but in a much simpler package. (Photo Courtesy Holley/Nitrous Oxide Systems)

I consider a billet-steel flywheel a must for any performance engine backed by a manual transmission. Under no circumstances should a cast-iron original be used. It's simply too risky. Centerforce presently produces the best Pontiac flywheel available on the aftermarket. It's SFI certified and fits and functions just like an original.

A flexplate connects the torque converter of an automatic transmission to the flywheel. Constructed of a sheet metal, the holes cut into it allow it to flex slightly, taking up torque converter movement at higher engine speeds. Originals were adequately made for stock applications, but the aftermarket unit from the TCI unit is much beefier. It's an excellent choice that's SFI certified and fits quite well.

the oxygen content increase, the result is a much more intense combustion event, which can improve engine performance. The nitrogen molecules are inert and sent out the tailpipe with the rest of the exhaust.

Many companies produce nitrous oxide kits designed for automotive use. They all include a large bottle located in the rear of the car and the appropriate plumbing and switches that control delivery into the engine.

While significant performance gains are possible, the use of nitrous oxide increases the amount of cylinder pressure and heat within the engine, and that can be tough on certain engine components. When using enough to increase horsepower by 100 or more, it's advisable to use forged connecting rods and pistons, and to gap the piston rings accordingly. If you plan to use nitrous oxide with your Pontiac, I highly advise discussing your goals with a professional Pontiac engine builder who specializes in nitrous oxide applications, such as Butler Performance or Ken's Speed & Machine Shop.

Flywheel and Flexplate

The quality of the flywheel or flexplate should be serious concerns in any build that takes performance beyond the stock level. Either unit bolts to the flange at the rear of the crankshaft and connects the engine to the transmission. Original units are around 40 years old, and have endured countless miles of use and heating/cooling cycles. You should strongly consider high-quality aftermarket units with SFI-certification for any high-performance build, as well as the use of high-quality fasteners such as those from ARP.

A flywheel provides a mating surface for the clutch disc and pressure plate assembly of a manual transmission. Originals were constructed of cast iron and minor stress cracks on the clutch surface from years of use are common. While those cracks are generally machined away during resurfacing, they can compromise the integrity of the flywheel. It can literally explode when heavily loaded, significantly damaging the engine and transmission, or even you!

I highly recommend a billet-steel crankshaft for any Pontiac V-8 backed by a manual transmission. Centerforce offers a very high quality flywheel that weighs 30 pounds, or about 10 less than a typical cast-iron original. It's available with a crankshaft bore measuring 2.5 inches or 2.75, for whichever crankshaft you have. It's an excellent piece that fits and functions like a stocker. It sells for about $350 and is compatible with 10.4- and 11-inch clutch disc and pressure plate assemblies. If you need a high-quality clutch kit, Centerforce offers several. I use its Dual Friction for my own Pontiac and gripping power is excellent and pedal effort is amazingly low.

Like a flywheel, a flexplate bolts to the flywheel flange and it connects the torque converter of an automatic transmission to the engine. While rather rigid, it gets its name because it's constantly flexing to allow for torque converter movement at higher engine speeds. Over time that flexing can cause stress cracks that can ultimately end in failure, which, like a flywheel, can damage you and/or your vehicle. TCI offers a heavy duty flexplate for the Pontiac V-8 that's much thicker than stock and is SFI certified. Selling for $250, it's an excellent choice for any application where an automatic transmission is used.

Water Pump

An engine's water pump circulates coolant throughout the engine. The original design worked suitably, but finding OE-replacement water pumps with the same quality and volume capacity as an original can be difficult. Flowkooler produces a high-quality replacement that features an aluminum housing but cast-iron impeller. It's an excellent choice for any high-performance build and is one I use on my own Pontiacs.

Circulating that coolant throughout the engine consumes a slight amount of engine power. Meziere offers a very high quality electric water pump that fits like an original, but is capable of moving 35 or 42 gpm, depending upon which you select. Powered by a 12-volt source, it doesn't consume any engine power in the process. With a rated life of 2,500 hours of operation, it can add benefit to any high-performance Pontiac V-8, but particularly those where extracting every last horsepower is important, or in instances where circulating coolant between rounds without the engine running can improve performance. Pricing starts at $400.

Starter

A stock Pontiac starter is designed to turn over a production-type Pontiac V-8 that operates on pump gas and uses original equipment. Significantly increasing the compression ratio, such as that required for competition engines, raises the amount of cylinder pressure the starter must work against when attempting to fire the engine. Combine that with a base timing setting that's relatively high and the engine will likely kick back against the starter, refusing to start. A similar situation occurs when the heat radiated by tubular headers causes the internal windings of the steel-cased starter to expand, which can causes the start to drag or simply "click" in protest.

When dealing with a high-performance engine, it's best to forego using any original Pontiac starter and opt for one of the many high-quality gear-reduction starters on the market today. It typically uses a high-speed motor to transmit power via to an offset gear to turn the engine over. The result is a compact unit that's more efficient, weighs less, and cranks the engine at a higher speed with more torque than a stock starter.

IMI Performance Products produces one of the best gear-reduction starters on the market today. It features a high-speed electric motor and a solenoid encased in aluminum to dissipate heat quicker than the stock unit, and it includes a vent tube to increase overall airflow through the unit. IMI offers several starters, and the difference is in cranking output. Pricing starts at about $250. It's an excellent product that installs easily and performs reliably in all types of applications and operating conditions.

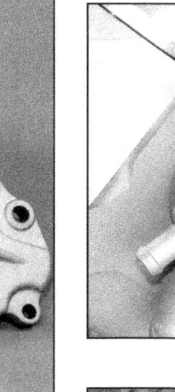

Flow Kooler water pumps are popular belt-driven units that install easily and function very well. It features a cast-aluminum housing and cast-iron impeller for a lightweight unit that operates efficiently and reliably. It's my first choice when selecting a stock-type water pump for my own builds.

Meziere produces two high-quality electric water pumps for Pontiac V-8s in a variety of colors. Its 35-gph unit is recommended for naturally aspirated applications producing up to 650 hp, and its 42-gph unit is recommended for applications that exceed that. Contact Meziere for its recommendation if you're running forced induction or heavy doses of nitrous oxide.

IMI offers an excellent mini starter for the Pontiac V-8 that has high-torque, gear-reduction starting power and aluminum construction for quick heat dissipation. It produces specific units for stock-type applications and others with increased cranking power for engines with very high compression ratios. I highly recommend such a starter for your Pontiac build.

CHAPTER 13

PERFORMANCE COMBINATIONS

A 414-inch Pontiac V-8 is capable of delivering strong horsepower and masssive amounts of usable torque while operating reliably on pump gas. This Butler Performance combo made 578 hp on 588 ft-lbs and operates on 89-octane fuel. It's extremely docile and sounds quite healthy.

When building a max-performance Pontiac V-8, each engine is almost always unique unto itself. Depending upon the racing class restrictions, chassis hurdles, intended performance level and usage, and budgetary concerns, very few max-performance efforts are exactly identical.

There is a wide variety of machine shops, engine assemblers, and engine builders that are quite capable of building a very strong Pontiac V-8—and many specialize in and race Pontiacs. The Source Guide provides contact information for companies that I am comfortable recommending.

I contacted a select number of Pontiac-specific engine builders and asked them to provide recommendations for specific max-performance engine builds. Some builds include a stock Pontiac block while others use an aftermarket block, some are naturally aspirated while others use forced induction or accommodate nitrous oxide, and some operate on commonly available pump gas while others require race gas. The only restriction that I imposed upon each builder was that the engine package must operate reliably and perform within a reasonable range of the estimated horsepower and torque values.

The max-performance combinations on the following pages can be considered guidelines to arrive at the supplied horsepower and torque estimations. While any number of the recommended components could be replaced with like items, performance or reliability could suffer. If your build deviates from these, I strongly suggest you consult with your builder to determine the effects that might have.

Butler Performance: 575+ hp, 474 ci, Pump Gas

Reliable operation on premium-grade pump gas and generous power amounts are quite possible when combining a stock Pontiac block and certain aftermarket components. With the significant cost of 91- to 93-octane fuel these days, combinations compatible with 89-octane gas are gaining popularity. With the assistance of Butler Performance, we built a 474 that provides more than 575 hp on mid-grade fuel. Horsepower and torque greater than 600 is possible when pushing compression toward 10.5:1, but it requires a steady diet of premium-grade fuel. If you plan on driving your Pontiac regularly, and want a low-maintenance design that operates detonation-free, this combination may be for you.

Short Block

Block Casting: Any 1970–1976 455-ci unit
Preparation: Bore and hone with torque plate
Crankshaft: Butler Performance, forged-steel, 4.25-inch stroke, 3.25-inch main journal, 2.20-inch rod journal
Preparation: Normal
Bore/Stroke: 4.212 x 4.25 inches
Displacement: 473.5 ci
Deck Height: zero
Pistons: Ross custom, forged aluminum, 4.211-inch diameter, flat top, 439.5 grams, 8.5-cc reliefs
Piston Pins: Ferrea No. P70028-8, forged alloy, 0.990-inch diameter, 138 grams
Piston Rings: Total Seal No. CR3455-65, plasma-moly, file fit
Piston Ring End Gap: .018 inch
Piston-to-Wall Clearance: .005 inch
Connecting Rods: Eagle Specialty Products, forged steel, H-beam, 6.800-inch length, 7/16-inch ARP bolts
Connecting Rod Side Clearance: .025 to .028 inch
Main Bearings: Federal Mogul No. 151M-STD main journal
Main Bearing Clearance: .003 inch
Rod Bearings: Federal Mogul 8-7200CH-STD rod journal
Rod Bearing Clearance: .003 inch
Main Caps: Pro Gram Engineering four-bolt
Main Cap Bolts/Studs: ARP main studs
Thrust Clearance: .006 to .009 inch
Windage Tray: Tomahawk
Oil Pump: Butler Pro-Series
Oil Pan: Canton No. 15-450, road-race
Oil Pump Driveshaft: Melling
Block Filler: Optional, 1 inch below frost plugs
Harmonic Damper: Pioneer Performance Street No. PBO-PB1056ST

Camshaft

Camshaft: Comp Cams custom hydraulic roller
Lifters: Comp Cams hydraulic roller
Camshaft Bearings: Federal Mogul
Bearing Clearance: Can easily rotate by hand
Duration at .050-inch: 242/248 degrees
Lobe Separation: 110 degrees
Valve Lift at 1.5: .540/.562 inch
Intake Centerline Installed at: 106.5 degrees
Pushrods: Comp Cams Hi-Tech
Rocker Studs: ARP 7/16 inch
Rocker Arms: Comp Cams roller

Cylinder Heads

Castings: Edelbrock Performer RPM
Preparation: Ported by Butler Performance
Combustion Chamber Volume: 90 cc
Compression Ratio: 9.95:1
Suggested Intake/Exhaust Airflow Flow at 28 inches: 312-cfm intake, 230-cfm exhaust at .600 inch lift
Intake Port Volume: Approximately 220 cc
Valves: Ferrea stainless steel
Valve Diameters: 2.11/1.77 inches
Valve and Seat Preparation: Multi-angle with 45-degree seats
Valveguides: Bronze, 0.530-inch diameter
Valve Seals: Viton
Valvesprings: Lunati No. 73100-16
Valvespring Pressure: 137 pounds seat, 338 pounds open
Valve Locks and Retainers: Comp Cams
Head Gaskets: Butler Performance
Cylinder Head Fasteners: ARP head bolt kit

Induction

Intake Manifold: Edelbrock Victor No. 2957, modified by Butler Performance
Preparation: Ported
Carburetor: Holley Ultra HP 850 cfm

Ignition

Distributor: MSD HEI
Distributor Gear: BOP Composite
Spark Plugs: NGK R5671A-7
Spark Plug Gap: .050 inch
Ignition Amplifier: MSD
Coil: MSD
Total Timing: 40 degrees
Suggested Fuel Octane: 89

Exhaust

Exhaust Headers: Doug's Header four-tube
Primary Tube Diameter: 1.75 inches
Collector Diameter: 3 inches

Projected Output

Peak Horsepower: 578 at 5,500 rpm
Peak Torque: 588 ft/lbs at 4,700 rpm

CHAPTER 13

Ken's Speed & Machine Shop: 700 hp, 535 ci, Pump Gas

Ken Keefer is a popular Pontiac engine builder located in Brooksville, Florida. He specializes in large-cube Pontiac V-8s that operate on commonly available pump-gas, and are capable of dominating the street classes at the drag strip. Through much trial and error, Ken has uncovered what it takes to make a Pontiac V-8 survive on rather large doses of nitrous oxide. The combination he provide operates naturally aspirated, but can easily produce more than 850 hp with a 175-shot of nitrous oxide.

Ken added, "This engine has produced quarter-mile times of 9.70 seconds at 136 mph in my 1976 Trans Am weighing 3,450 pounds with stock suspension. It's well into the 8s with a 175-hp shot of nitrous."

Short Block

Block Casting: IA-2 from AllPontiac.com
Preparation: Hone to tolerance with aluminum torque plate
Crankshaft: Forged 4340-steel from RPM International or SCAT
Preparation: Check that all journals are round, and balance
Bore/Stroke: 4.35 x 4.5 inches
Displacement: 535 ci
Deck Height: 10.3 inches
Pistons: B.R.C. Custom left .005 inch below deck surface
Piston Rings: Hastings Pro Race
Piston Ring End Gap: .020-inch top, .024-inch second (with nitrous, .024-inch top, .028-inch second)
Piston-to-Wall Clearance: .0035 to .0045 inch
Connecting Rods: Eagle or SCAT 4340 H-beam, 6.7-inch length
Connecting Rod Side Clearance: .018 to .020 inch
Main Bearings: Federal-Mogul No. 113M Performance with 3/4 oil groove
Main Bearing Clearance: .002 to .0025 inch
Rod Bearings: Clevite 77 Race No. CB743HN
Rod Bearing Clearance: .002 inch
Main Caps: four-bolt splayed caps included with IA-2 block
Main Cap Bolts/Studs: ARP studs
Thrust Clearance: .005 to .007
Windage Tray: Short factory or Pacific Performance Racing tray
Crank Scraper: Yes
Oil Pump: M54D with .060-inch shim under relief spring
Oil Pan: Canton
Oil Pump Driveshaft: Melling No. IS-54A
Block Filler: None
Harmonic Damper: Pioneer SFI
Rear Seal: Best Graphite Rope Seal

Camshaft

Camshaft: Comp Cams hydraulic roller
Lifters: Comp Cams hydraulic roller
Camshaft Bearings: ACL or Durabond P-4
Bearing Clearance: Should be able to spin with finger and thumb
Duration at .050-inch: 282/288 degrees
Lobe Separation: 110 degrees
Valve Lift at 1.65: .648 inch
Intake Centerline Installed at: 106 to 107 degrees
Pushrods: Comp Cams Hi-Tech
Rocker Studs: ARP 7/16 inch
Rocker Arms: Harland Sharp 1.65:1 ratio

Cylinder Heads

Castings: Edelbrock Performer RPM
Preparation: CNC ported by SD Performance
Combustion Chamber Volume: 88 cc
Compression Ratio: 10.8:1
Suggested Intake/Exhaust Airflow Flow at 28 inches: 330-cfm intake, 250-cfm exhaust
Valves: Manley stainless steel
Valve Diameters: 2.2/1.74 inches
Valve and Seat Preparation: 30-degree back-cut on intake valves, 5-angle intake seat, 4-angle exhaust seat
Valveguides: Edelbrock bronze
Valve Seals: SI viton
Valvesprings: Comp Cams beehive-type
Valvespring Pressure: 145 pounds seated, 405 pounds open
Valve Locks and Retainers: Comp Cams
Head Gaskets: Cometic, .051 inch thick
Cylinder Head Fasteners: ARP studs

Induction

Intake Manifold: Victor 4500 series
Preparation: ported
Carburetor: 1,100-cfm Pro Systems Holley

Ignition

Distributor Type: MSD Pro-billet
Distributor Gear: BOP Composite
Crank Trigger: No
Spark Plugs: NGK R-5671A-8
Spark Plug Gap: .040-inch gap
Ignition Amplifier: MSD Digital Box
Coil: MSD Power Coil
Range of Initial Timing: 12 to 16 degrees
Total Timing: 36 to 38 degrees
RPM Total Should be Reached by: 3,000 rpm
Suggested Fuel Octane: 93

Exhaust

Exhaust Headers: Hooker No. 4202
Primary Tube Diameter: 2 inches
Collector Diameter: 3.5 inches
Exhaust Tube Diameter (if required by race track): 3.5 inches
Muffler: DynoMax Bullet
Exhaust Crossover: None

Projected Output

Estimated Horsepower: 695 hp
Estimated Torque: 680 ft-lbs

SD Performance: 900 hp, 535 ci, Race Gas

Dave Bisschop takes a very serious approach toward max-performance Pontiacs. He has been building and racing Pontiacs for several decades, and is always looking for new ways to extract even more performance. Some of the quickest Pontiacs in the country feature cylinder heads that have been ported by his company, SD Performance in Chilliwack, British Columbia. Dave is considered to be among the very best Pontiac builders in the country today and the combinations he assembles delivers the performance he claims.

Short Block

Block Casting: MR-1
Preparation: Check mainline for size, bore, torque plate hone, deck
Crankshaft: Eagle forged 4340-steel
Preparation: Inspect, measure, polish and clean
Bore/Stroke: 4.35 x 4.5 inches
Displacement: 535 ci
Deck Height: 10.23 inches
Pistons: Ross custom flat top
Piston Rings: Plasma Moly file fit, Napier second,
Piston Ring End Gap: .020-inch top, .024-inch second
Piston-to-Wall Clearance: .005 to .006 inch
Connecting Rods: Callies forged, 6.66-inch length
Connecting Rod Side Clearance: .028 to .030 inch
Main Bearings: Clevite 77 standard,
Main Bearing Clearance: .0022 to .0028 inch
Rod Bearings: Federal-Mogul competition
Rod Bearing Clearance: .0025 to .0030 inch
Main Caps: Pro-gram Engineering four-bolt splayed (included with MR-1 block)
Main Cap Bolts/Studs: ARP studs
Thrust Clearance: .006 to .009 inch
Windage Tray: Canton
Oil Pump: Melling M54D, disassemble, debur, clean, measure bottom clearance
Oil Pan: Canton Road Race
Oil Pump Driveshaft: Melling
Block Filler: No
Harmonic Damper: ATI

Camshaft

Camshaft: LSM solid roller, revised firing order
Lifters: Crower solid roller with high-pressure oiling
Camshaft Bearings: Durabond
Bearing Clearance: .0012 to .0017 inch
Duration at .050-inch: 276/286 degrees
Lobe Separation: 114 degrees
Valve Lift at 1.5: .681 inch
Intake Centerline Installed at: 110 to 111 degrees
Pushrods: Smith Bros., 3/8 inch thick, .145-inch wall
Rocker Studs: No
Rocker Arms: Crower or T&D shaft rockers, 1.7:1 ratio

Cylinder Heads

Castings: Edelbrock Performer RPM
Preparation: Offset intake pushrod holes, welded/machined semi-heart shape chambers, CNC porting
Combustion Chamber Volume: 68 cc
Compression Ratio: 14:1
Suggested Intake/Exhaust Airflow Flow at 28 inches: 360-cfm intake, 250-cfm exhaust
Valves: Manley custom stainless steel
Valve Diameters: 2.25/1.74 inches
Valve and Seat Preparation: 30-degree back-cut on intake valves, 5-angle intake seat, 4-angle exhaust seat
Valveguides: Edelbrock bronze
Valve Seals: SI viton
Valvesprings: Comp No. 98745 dual with damper
Valvespring Pressure: 250 pounds seat, 690 pounds open
Valve Locks and Retainers: Titanium 10 degree
Head Gaskets: Cometic, .040 inch thick
Cylinder Head Fasteners: ARP studs

Induction

Intake Manifold: Edelbrock Victor 4500
Preparation: CNC ported runners, CNC-machined plenum and hand finishing
Carburetor: Pro Systems Dominator 1,150+ cfm

Ignition

Distributor Type: MSD Billet
Distributor Gear: BOP composite
Crank Trigger: None
Spark Plugs: Autolite No. 3924
Spark Plug Gap: .035 to .040 inch
Ignition Amplifier: MSD
Coil: MSD
Total Timing: Locked at 36 to 38 degrees
Suggested Fuel Octane: 108 to 110

Exhaust

Exhaust Headers: Dougs, Hooker or Hedman
Primary Tube Diameter: 2 inches
Secondary Tube Diameter: 3.5 inches
Exhaust Tube Diameter (if required by race track): 3.5 to 4 inches
Mufflers: Spintech
Exhaust Crossover: X-type crossover

Projected Output

Estimated Horsepower: 900 hp
Estimated Torque: 750 ft-lbs

Butler Performance: 1,000+ hp, Turbocharged 505 ci, Pump Gas

Butler Performance in Leoma, Tennessee, has been building fast Pontiacs for a number of years and sons David and Rodney have picked up where their father, Jim, left off. David Butler is among the most knowledgeable Pontiac builders around today and his approach toward any type of Pontiac performance can be backed up with successful results. With the ability to produce a stout performance engine with a stock appearance or thousands of horsepower from a boosted application, Butler Performance can assist you with all your Pontiac needs.

Short Block

Block Casting: IA-II high-nickel block, extra thick deck
Preparation: Pressure tested, sonic tested, deburred, bored and honed with torque plate, deck squared
Crankshaft: SCAT 4340-forged steel
Preparation: Inspect and clean
Bore/Stroke: 4.35 x 4.25 inches
Displacement: 505 ci
Deck Height: 10.23 inches
Pistons: Ross, custom to Butler specs
Piston Rings: Total Seal TNT file fit
Piston Ring End Gap: .026 to .030 inch
Piston-to-Wall Clearance: .006 to .0065 inch
Connecting Rods: Oliver billet steel, 6.7 inches, fit pin and check sizes
Connecting Rod Side Clearance: .025 to .028 inch
Main Bearings: Federal-Mogul Race Series
Main Bearing Clearance: .003 inch
Rod Bearings: Federal-Mogul Race Series
Rod Bearing Clearance: .003 inch
Main Caps: Standard IA-11 four-bolt billet main caps, splayed centers
Main Cap Bolts/Studs: ARP studs
Thrust Clearance: .006 to .009 inch
Windage Tray: Option
Crank Scraper: Yes, Butler Performance
Oil Pump: Butler Performance Pro pump or Titan billet gerotor pump
Oil Pan: Canton or billet fabrication, custom aluminum
Oil Pump Driveshaft: Melling
Block Filler: No
Harmonic Damper: ATI Super Damper

Camshaft

Camshaft: Comp Cams, hydraulic roller
Lifters: Comp Cams
Camshaft Bearings: Federal-Mogul
Bearing Clearance: Can easily rotate by hand
Duration at .050-inch: 254/246 degrees
Lobe Separation: 115 to 116 degrees
Valve Lift at 1.5: .620 inch
Intake Centerline Installed at: 113 to 114 degrees
Pushrods: 3/8-inch diameter
Rocker Arms: T&D rocker shaft, 1.65:1 ratio

Cylinder Heads

Castings: Edelbrock Victor High Port
Preparation: Ported
Combustion Chamber Volume: 87 cc
Compression Ratio: 8:1 for pump gas
Suggested Intake/Exhaust Airflow Flow at 28 inches: 400-cfm intake, 270-cfm exhaust
Valves: Ferrea stainless steel
Valve Diameters: 2.19/1.77 inches
Valve and Seat Preparation: 3-angle competition valve job, 45-degree seats
Valveguides: Edelbrock bronze
Valve Seals: Viton
Valvesprings: Comp Cams, 165 to 185 pounds closed, 380 pounds open
Valve Locks and Retainers: Chrome-moly locks and Titanium retainers
Head Gaskets: Cometic
Cylinder Head Fasteners: ARP studs

Induction

Intake Manifold: Edelbrock Super Victor EFI or Hogan custom sheet metal
Preparation: Ported
Carburetor: FAST XFI, electronic fuel injection

Power Adder

Turbo: Precision Turbo 88 mm
Suggested Boost: 12 psi
Extra Components: Tial waste gate, Tial blow-off valve, Precision Turbo intercooler

Ignition

Distributor Type: MSD Pro-Billet
Distributor Gear: BOP Composite
Crank Trigger: MSD
Spark Plugs: NGK
Spark Plug Gap: .028 to .030 inch
Ignition Amplifier: MSD
Coil: MSD
Range of Total Timing: Controlled by EFI software program
Suggested Fuel Octane: 93

Exhaust

Exhaust Headers: Custom mild steel or optional stainless steel
Primary Tube Diameter: 1 7/8 inches

Projected Output

Estimated Horsepower: 950 to 1,200 hp depending upon boost
Estimated Torque: 1,000 ft-lbs

(Photo Courtesy Don Keefe)

PERFORMANCE COMBINATIONS

Butler Performance: 1,100+ hp, 541 ci, Race Gas

Short Block

Block Casting: IA-II high-nickel cast iron, or All Pontiac cast-aluminum
Preparation: Pressure tested, sonic tested, deburred, bored and honed with torque plate, deck squared
Crankshaft: SCAT 4340-forged steel or Crower billet steel
Preparation: Inspect and clean
Bore/Stroke: 4.375 x 4.5 inches
Displacement: 541ci
Deck Height: 10.22 inches
Pistons: Ross, custom to Butler specs
Piston Rings: Total Seal No. CS-9190-130, AP stainless top ring, Napier 2nd ring
Piston Ring End Gap: .024 inch
Piston-to-Wall Clearance: .006 inch
Connecting Rods: Oliver billet steel or GRP billet aluminum, 6.7 inches, fit pin and check sizes
Connecting Rod Side Clearance: .025 to .028 inch
Main Bearings: Federal-Mogul Race Series
Main Bearing Clearance: .003 inch
Rod Bearings: Federal-Mogul Race Series
Rod Bearing Clearance: .003 inch
Main Caps: Standard IA-11 four-bolt billet main caps, splayed centers
Main Cap Bolts/Studs: ARP Studs
Thrust Clearance: .006 to .009 inch
Windage Tray: Built in to oil pan, crank scraper
Oil Pump: Butler Pro Series or Titan billet gerotor type
Oil Pan: Canton steel or custom billet aluminum
Oil Pump Driveshaft: Melling
Block Filler: No
Harmonic Damper: ATI Super Damper

Camshaft

Camshaft: Comp Cams custom solid roller
Lifters: Crower Severe duty, offset intake
Camshaft Bearings: 50-mm roller
Bearing Clearance: Should turn easily by hand
Duration at .050-inch: 289/300 degrees
Lobe Separation: 114 degrees
Valve Lift at 1.5: .890/.850 inch
Intake Centerline Installed at: 112 degrees
Pushrods: 3/8- or 7/16-inch diameter with .120-inch wall
Rocker Studs: None
Rocker Arms: T&D Shaft Rockers, 1.7:1 to 1.8:1 ratio

Cylinder Heads

Castings: Edelbrock Victor High Port
Preparation: Fully CNC ported and CNC chambers
Combustion Chamber Volume: 60 cc
Compression Ratio: 15.2:1
Suggested Intake/Exhaust Airflow Flow at 28 inches: 440-cfm intake, 300-cfm exhaust
Valves: Ferrea Titanium
Valve Diameters: 2.3/1.7 inches
Valve and Seat Preparation: Competition valve job with 55-degree seats
Valveguides: Bronze
Valve Seals: Viton
Valvesprings: Manley or PAC
Valvespring Pressure: 300 pounds seat, 900+ pounds open
Valve Locks and Retainers: Titanium
Head Gaskets: Cometic, .040 inch thick
Cylinder Head Fasteners: ARP studs

Induction

Intake Manifold: BOP Engineering single 4-barrel (or custom Hogan aluminum dual 4-barrel)
Preparation: Ported
Carburetor: Single 1,250-cfm Holley (or dual 1,150-cfm Holleys)

Ignition

Distributor Type: MSD Pro Billet
Distributor Gear: BOP composite
Crank Trigger: MSD
Spark Plugs: NGK
Spark Plug Gap: .030 to .040 inch
Ignition Amplifier: MSD Digital 7
Coil: MSD
Total Timing: Locked at 32 to 36 degrees
Suggested Fuel Octane: 116

Exhaust

Exhaust Headers: Custom
Primary Tube Diameter: 2.125 inches
Collector Diameter: 4 to 4.5 inches

Projected Output

Estimated Horsepower: 1,100+ hp
Estimated Torque: 900 ft-lbs

CHAPTER 13

Kauffman Racing Equipment: 1,200 hp, 535 ci, Race Gas

Kauffman Racing Equipment (KRE) in Glenmont, Ohio, is among the very best in the today's Pontiac performance market. Since introducing its own cast-aluminum d-port cylinder head for Pontiac applications a few years ago, KRE has gone on to develop cast-iron and -aluminum large-bore Pontiac blocks and high-performance cylinder heads known as the High Port and Warp 6. Jeff Kauffman provides a very high performance combination using his company's newest high-performance cylinder head.

Jeff added, "This would be what I consider our full-out race engine. It is designed to make maximum horsepower when compared to any other Pontiac engine combo and to be as light as possible."

Short Block

Block Casting: K&M Performance MR-1A aluminum
Preparation: Bore and hone, install bronze lifter bushings
Crankshaft: SCAT forged 4340-steel
Preparation: Measure specs and balance
Bore/Stroke: 4.35 x 4.5 inches
Displacement: 535 ci
Deck Height: 10.24 inches
Pistons: Ross forged-aluminum custom flat top
Piston Rings: Total Seal custom, file fit
Piston Ring End Gap: .020-inch top, .020-inch second
Piston-to-Wall Clearance: .006 to .010 inch
Connecting Rods: GRP billet aluminum, 6.7-inch length
Connecting Rod Side Clearance: minimum of .018 inch
Main Bearings: Federal-Mogul
Main Bearing Clearance: .002 to .004 inch
Rod Bearings: Federal-Mogul
Rod Bearing Clearance: .003 to .0035 inch
Main Caps: four-bolt caps standard with MR-1A block
Main Cap Bolts/Studs: ARP studs
Thrust Clearance: .007 inch
Windage Tray: Yes, built into oil pan
Oil Pump: Nutter Dry Sump, 4 or 5 stage
Oil Pan: Stef's custom made to KRE spec for dry sump
Oil Pump Driveshaft: None
Block Filler: No
Harmonic Damper: Romac SFI steel/aluminum
Vacuum pump: Yes

Camshaft

Camshaft: LSM 55-mm custom solid roller
Lifters: Crower solid roller, .937-inch diameter
Camshaft Bearings: 55-mm roller bearings
Bearing Clearance: .002 to .003 inch
Duration at .050 inch: 280/310 degrees
Lobe Separation: 114 to 118 degrees
Valve Lift with 1.5:1: .840 inch
Intake Centerline Installed at: 110 to 115 degrees
Pushrods: Custom .5-inch diameter
Rocker Studs: None
Rocker Arms: KRE/T&D shaft rockers, 1.7:1 ratio

Cylinder Heads

Castings: KRE Warp 6
Preparation: CNC ported
Combustion Chamber Volume: 60 cc
Compression Ratio: 15.5:1
Suggested Intake/Exhaust Airflow Flow at 28 inches: 500-cfm intake, 400-cfm exhaust
Valves: Ferrea Titanium
Valve Diameters: 2.4/1.85 inches
Valve and Seat Preparation: Custom 5-angle valve job
Valveguides: Bronze
Valve Seals: PC seals
Valvesprings: PSI dual springs
Valvespring Pressure: 400 pounds seat, 1,000 pounds open
Valve Locks and Retainers: 10-degree Titanium
Head Gaskets: Cometic, .051 inch
Cylinder Head Fasteners: ARP Studs

Induction

Intake Manifold: KRE custom with sheet-metal plenum and billet runners, dual carburetors
Preparation: None
Carburetor: Two Holley Dominators, 1,250-cfm each, 1-inch HVH carburetor spacers

Ignition

Distributor Type: MSD billet
Distributor Gear: MSD bronze
Crank Trigger: MSD
Spark Plugs: NGK R5671A-10
Spark Plug Gap: .035 inch
Ignition Amplifier: MSD Digital 7
Coil: MSD HVC
Total Timing: Locked at 33 degrees
Suggested Fuel Octane: 116

Exhaust

Exhaust Headers: Custom KRE
Primary Tube Diameter: 2.25 inches
Secondary Tube Diameter: none

Projected Output

Estimated Horsepower: 1,200 hp
Estimated Torque: 820 ft-lbs

Source Guide

All Pontiac.com
11010 Trade Road
Richmond, VA 23236
www.allpontiac.com
804-794-6777

Ames Performance Engineering
10 Pontiac Drive
PO Box 572
Spofford, NH 03462
www.amesperf.com
800-421-2637

ARP Bolts
1863 Eastman Avenue
Ventura, CA 93003
www.arp-bolts.com
800-826-3045

Best Gasket Company
11558 E Washington Boulevard, Suite F
Whittier, CA 90606
www.bestgasket.com
888-333-2378

BHJ Products
37530 Enterprise Court
Newark, CA 94560
www.bhjproducts.com
510-797-6780

Bob Davis Distributors
660 Tamburlaine Cove
Collierville, TN 38017
901-412-4414

BOP Engineering
N3651 Schmidt Road
Jefferson, WI 53549
www.bopengineering.com
920-674-6058

Brad Penn
77 North Kendall Avenue
Bradford, PA 16701
www.penngrade1.com
814-368-1200

Butler Performance
2336 Hwy 43 S
Leoma, TN 38468
www.butlerperformance.com
866-762-7527

Canton Racing Products
232 Branford Road
North Branford, CT 06471
203-481-9460
www.cantonracingproducts.com

Carrillo Rods
1902 McGaw Avenue
Irvine, CA 92614
949-567-9000
www.cp-carrillo.com

Centerforce Clutches
2266 Crosswind Drive
Prescott, AZ 86301
www.centerforce.com
928-771-8422

Central Virginia Machine Service
606B Second Street NW
Burkeville, VA 23922
www.centralvirginiamachine.com
434-767-9915

Cliff's High Performance
20579 Berry Road
Mount Vernon, OH 43050
www.cliffshighperformance.com
740-397-2921

Cometic Gaskets
8090 Auburn Road
Concord, OH 44077
www.cometic.com
800-752-9850

SOURCE GUIDE

Comp Cams
3406 Democrat Road
Memphis, TN 38118
www.compcams.com
800-999-0853

Crane Cams
1640 Mason Avenue, Unit 180
Daytona Beach, FL 32117
www.cranecams.com
866-350-5120

Crower Cams and Equipment
6180 Business Center Court
San Diego, CA 92154
www.crower.com
619-661-6477

Dave's Small Body HEI's
24 Buffalo Lane
Yerington, NV 89447
www.davessmallbodyheis.com
775-722-3294

DCI Motorsports
3573 B Gilchrist Road
Mogadore, OH 44260
www.dcimotorsports.com
330-628-3354

Dura-Bond Bearings
3200 Arrowhead Drive
Carson City, NV 89706
www.dura-bondbearing.com
800-227-8360

Eagle Specialty Products
8530 Aaron Lane
Southaven, MS 38671
www.eaglerod.com
662-796-7373

Edelbrock
2700 California Street
Torrance, CA 90503
www.edelbrock.com
310-781-2222

Federal-Mogul
26555 Northwestern Highway
Southfield, MI 48033
www.federalmogul.com
248-354-7700

Ferrea Valves
2600 NW 55 Court, Suite # 234
Ft. Lauderdale, FL 33309
www.ferrea.com
888-733-2505

Flowkooler
500 Linne Road, Unit I
Paso Robles, CA 93446
www.flowkooler.com
802-239-2501

GRP Connecting Rods
333 W 48th Avenue
Denver, CO 80216
www.grpconrods.com
303-935-7565

Hard Blok
PO Box 1274
Brentwood, TN 37024
www.hardblok.com
865-457-0509

Hedman Street Hedders
12438 Putman Street
Whittier, CA 90602
www.hedman.com
562-921-0404

High Performance Pontiac Magazine
PO Box 420235
Palm Coast, FL 32142
www.highperformancepontiac.com
800-568-6130

Holley Performance Carburetors
1801 Russellville Road
Bowling Green, KY 42101
www.holley.com
270-782-2900

H-O Enterprises
8780 Bajado Court
Rancho Cucamonga, CA 91730
www.hoenterprises.com
909-980-1451

Jim Taylor Engine Service
120 S 5th Street
Phillipsburg, NJ 08865
908-213-3456

IMI Performance Products
13423 Lambert Road
Whittier, CA 90605
www.hitorque.com
562-907-9400

Kauffman Racing Equipment
22280 Temple Road
Glenmont, OH 44628
www.krepower.com
740-599-5000

Keith Black Pistons
1040 Corbett Street
Carson City, NV 89706
www.kb-silvolite.com
800-648-7970

Ken's Speed & Machine Shop
512 S Main Street
Brooksville, FL 34601
www.pontiacdude.cc
352-796-8800

Len Williams Auto Machine
739 S 83rd East Avenue
Tulsa, OK 74112
www.lenwilliamsautomachine.com
918-836-7583

Mahle Clevite
1240 Eisenhower Place
Ann Arbor, MI 48108
www.mahleclevite.com
734-975-4777

SOURCE GUIDE

Melling Engine Parts
2620 Saradan Drive
Jackson, MI 49204
www.melling.com
517-787-8172

Merrick Motors
PO Box 2422
Longmont, CO 80502
www.pontiacengines.net
303-776-0877

Meziere
220 S Hale Avenue
Escondido, CA 92029
www.meziere.com
800-208-1755

Mike's TriPowers
6513 N Fox Chapel Trail
Edwards, IL 61528
www.pontiactripower.com
309-360-6385

Milodon
2250 Agate Court
Simi Valley, CA 93065
www.milodon.com
805-577-5950

Moldex Crankshaft Company
12255 Wormer Road
Redford, MI 48239
www.moldexcrankshaft.com
313-387-6099

Mr. Gasket
10601 Memphis Avenue, # 12
Cleveland, OH 44144
www.mrgasket.com
216-688-8300

MSD Ignition
1350 Pullman Drive, Dock No. 14
El Paso, TX 79936
www.msdignition.com
915-857-5200

Ohio Crankshaft Company
5453 SR 49 S
Greenville, OH 45331
www.ohiocrank.com
937548-7113

Oliver Racing Parts
1025 Clancy Avenue NE
Grand Rapids, MI 49503
www.oliverconnectingrods.com
616-451-8333

Pacific Performance Racing
264 E 22nd Street
San Pedro, CA 90731
www.pacificperformanceracing.com
310-823-4596

Performance Distributors
2699 Barris Drive
Memphis, TN 38132
www.performancedistributors.com
901-396-5782

Performance Trends
PO Box 530164
Livonia, MI 48153
www.performancetrends.com
248-473-9230

Performance Years
2705 Clemens Road, Building 105A
Hatfield, PA 19440
www.performanceyears.com
800-542-7278

Pertronix Performance Products
440 E Arrow Highway
San Dimas, CA 91773
www.pertronix.com
909-599-5955

PHS Automotive Services
PO Box 183251
Shelby Township, MI 48318
www.phs-online.com
586-781-5164

Pontiac Oakland Club International
PO Box 68
Maple Plain, MN 55359
www.poci.org
877-368-3454

Proform Parts
PO Box 306
Roseville, MI 48066
www.proformparts.com
586-774-2500

Pro-Gram Engineering
475 5th Street NE
Barberton, OH 44203
www.pro-gram.com
330-745-1004

Pypes Exhaust
2705 Clemens Road, Building 105A
Hatfield, PA 19440
www.pypesexhaust.com
800-421-3890

Ram Air Restorations
17861 W Pond Ridge Circle
Gurnee, IL 60031
www.ramairrestoration.com
800-421-8455

RobbMc Performance Products
2960 Cameron Court
Carson City, NV 89706
www.robbmcperformance.com
775-885-7411

RPM International
16313 Arthur Street
Cerritos CA 90703
www.racingpartsmaximum.com
562-926-9188

SD Performance
44408 Vedder Mountain Road
Chilliwack, BC Canada V2R 4C4
www.sdperformance.com
604-392-2211

SOURCE GUIDE

SI Valves
4477 Shopping Lane
Simi Valley, CA 93063
www.sivalves.com
800-564-8258

Spotts Performance
31 N Maple Avenue
Hatfield, PA 19440
www.spottsperformance.com
215-362-2336

SuperFlow Technology Group
4747 Centennial Bloulevard
Colorado Springs, CO 80919
www.superflow.com
800-471-7701

Taylor Cable Products
301 Highgrove Road
Grandview, MO 64030
www.taylorvertex.com
816-765-5011

TCI Auto
151 Industrial Drive
Ashland, MS 38603
www.tciauto.com
888-776-9824

The Carburetor Shop LLC
204 E 15th Street
Eldon, MO 65026
www.thecarburetorshop.com
573-392-7378

Tin Indian Performance
PO Box 1162
Uniontown, Ohio 44685
www.tinindianperformance.com
330-699-1358

Total Seal
22642 N 15th Avenue
Phoenix, AZ 85027
www.totalseal.com
800-874-2753

Wilhite Automotive
200 W Washington Street
Derby, KS 67037
www.wilhiteauto.com
316-788-0514

Willard Auto Machine
7620 N 96th Street
Omaha, NE 68122
www.wampowered.com
402-573-8984

Wilson Manifolds
4700 NE 11 Avenue
Fort Lauderdale, FL 33334
www.wilsonmanifolds.com
954-771-6216

Year One
PO Box 521
Braselton, GA 30517
www.yearone.com
800-932-7663